Materials and Surface Engineering in Tribology

Materials and Surface Engineering in Tribology

Jamal Takadoum

First published in France in 2007 by Hermes Science/Lavoisier entitles: "Matériaux et surfaces en tribologie"
First published in Great Britain and the United States in 2008 by ISTE Ltd and John Wiley & Sons, Inc.
Translated from the French by Véronique Béguin

Apart from any fair dealing for the purposes of research or private study, or criticism or review, as permitted under the Copyright, Designs and Patents Act 1988, this publication may only be reproduced, stored or transmitted, in any form or by any means, with the prior permission in writing of the publishers, or in the case of reprographic reproduction in accordance with the terms and licenses issued by the CLA. Enquiries concerning reproduction outside these terms should be sent to the publishers at the undermentioned address:

ISTE Ltd
27-37 St George's Road
London SW19 4EU
UK

www.iste.co.uk

John Wiley & Sons, Inc.
111 River Street
Hoboken, NJ 07030
USA

www.wiley.com

© ISTE Ltd, 2008
© LAVOISIER, 2007

The rights of Jamal Takadoum to be identified as the authors of this work have been asserted by them in accordance with the Copyright, Designs and Patents Act 1988.

Library of Congress Cataloging-in-Publication Data

Takadoum, Jamal.
  Materials and surface engineering in tribology / Jamal Takadoum ; translated from the French by Véronique Béguin.
    p. cm.
  "Matériaux et surfaces en tribologie."
  Includes bibliographical references and index.
  ISBN 978-1-84821-067-7
  1. Tribology. 2. Surfaces (Technology). 3. Mechanical wear. I. Title.
  TJ1075.T3113 2008
  621.8'9--dc22
                                                                                2008037286

British Library Cataloguing-in-Publication Data
A CIP record for this book is available from the British Library
ISBN: 978-1-84821-067-7

Printed and bound in Great Britain by CPI Antony Rowe, Chippenham, Wiltshire.

FSC
Mixed Sources
Product group from well-managed
forests and other controlled sources

Cert no. SGS-COC-2953
www.fsc.org
© 1996 Forest Stewardship Council

*In memory of Professor Pierre-Gilles de Gennes
who honored me with his foreword for the original edition of this work.*

*A tribute to the genius of two great scientists of the 11th and 12th centuries:
Ibn Al-Haytham, the optical physicist, and Al-Jazari, the mechanical engineer.*

Table of Contents

Foreword xi

Preface xiii

Chapter 1. Surfaces 1
    1.1. Introduction. 1
    1.2. The surface state 2
        1.2.1. Structural state of a surface. 2
        1.2.2. Topographic state of a surface 3
            1.2.2.1. Atomic-scale topographic state 4
            1.2.2.2. Micrometer-scale topographic state. 5
            1.2.2.3. Experimental techniques 9
        1.2.3. Surface energy 18
            1.2.3.1 Surface energy measurements 21
        1.2.4. Mechanical state of a surface 24
            1.2.4.1. Hardness. 24
            1.2.4.2. Young's modulus 27
            1.2.4.3. Nano-indentation. 30
            1.2.4.4. Fracture toughness 34
            1.2.4.5. Residual stresses 37
        1.2.5. Chemical composition of a surface 43
            1.2.5.1. Energy dispersive X-ray analysis 44
            1.2.5.2. X-ray photoelectron spectroscopy. 45
            1.2.5.3. Auger electron spectroscopy 45
            1.2.5.4. Glow discharge optical emission spectroscopy 45
            1.2.5.5. Rutherford backscattering spectroscopy 46
            1.2.5.6. Secondary ion mass spectroscopy 46
            1.2.5.7. Infrared spectrometry 47

Chapter 2. Tribology 49

2.1. Introduction. 49
2.2. Elements of solid mechanics. 50
   2.2.1. The stress vector 50
   2.2.2. The stress tensor 50
   2.2.3. Yield criteria 52
      2.2.3.1. The Tresca criterion 52
      2.2.3.2. The von Mises criterion 53
2.3. Elements of contact mechanics 53
   2.3.1. Hertz contact theory 53
   2.3.2. The contact area 57
   2.3.3. Plastification of asperities 59
   2.3.4. Adhesive contact. 59
2.4. Friction 62
   2.4.1. The coefficient of friction 62
   2.4.2. Tribometers. 66
   2.4.3. Laws and theories of friction. 68
2.5. Nanotribology 71
   2.5.1. Surface forces 71
      2.5.1.1. Electrostatic forces. 71
      2.5.1.2. Capillary forces. 71
      2.5.1.3. Van der Waals forces 72
   2.5.2. Surface forces measurements 75
      2.5.2.1. The surface forces apparatus (SFA). 75
      2.5.2.2. The atomic force microscope (AFM) 76
      2.5.2.3. Application: surface forces and micromanipulation 79
   2.5.3. Nanofriction 81
2.6. Wear. 82
   2.6.1. The different forms of wear 83
      2.6.1.1. Adhesive wear. 83
      2.6.1.2. Abrasive wear. 83
      2.6.1.3. Fatigue wear 83
      2.6.1.4. Tribochemical wear 83
   2.6.2. Wear maps 84
   2.6.3. Interface tribology: third body concept. 86
   2.6.4. The PV product. 88
2.7. Lubrication 89
   2.7.1. Oils. 89
      2.7.1.1. The notion of viscosity 89
      2.7.1.2. The viscosity index and the SAE standard. 90
      2.7.1.3. The Stribeck curve. 91
      2.7.1.4. The different types of oils. 92

2.7.1.5. Greases	94
2.7.1.6. Anti-friction materials	94
2.8. Wear-corrosion: tribocorrosion and erosion-corrosion	98
2.8.1. Tribocorrosion	99
2.8.2. Erosion-corrosion	106

Chapter 3. Materials for Tribology 109

3.1. Introduction	109
3.2. Bulk materials	110
3.2.1. Metallic materials	110
3.2.1.1. Iron-based alloys	110
3.2.1.2. Superalloys	112
3.2.1.3. Copper-based alloys	113
3.2.2. Polymers	113
3.2.2.1. High-density polyethylene	115
3.2.2.2. Fluorinated polymers	115
3.2.2.3. Polyacetal (polyoxymethylene: POM) and polyamide	116
3.2.2.4. Polyimide	116
3.2.2.5. Polyetheretherketone (PEEK)	116
3.2.2.6. Friction and wear of polymers	117
3.2.2.7. Surface treatment of polymers	118
3.2.3. Composites	120
3.2.3.1. Friction materials	121
3.2.4. Ceramics	121
3.2.4.1. Friction and wear of ceramics	126
3.2.5. Cermets	131
3.2.5.1. Tungsten-carbide (WC)-based cermets	132
3.2.5.2. Other carbide-based cermets	133
3.3. Surface treatments and coatings	133
3.3.1. Conversion techniques	134
3.3.1.1. Anodic oxidation	134
3.3.1.2. Ion implantation	137
3.3.1.3. Ion beam mixing	142
3.3.1.4. Thermochemical treatment	143
3.3.1.5. Transformation hardening	143
3.3.1.6. Mechanical treatment	144
3.3.2. Deposition techniques	146
3.3.2.1. Thermal projection techniques	147
3.3.2.2. Liquid-phase deposition techniques	148
3.3.2.3. Vapor-phase deposition techniques	154
3.4. Hard anti-wear and decorative coatings	166

  3.4.1. Hard anti-wear coatings. 166
    3.4.1.1. Transition metal nitrides 166
    3.4.1.2. Carbon-based films 171
    3.4.1.3. The role of the substrate. 173
  3.4.2. Decorative coatings 174
 3.5. Characterization of coatings : hardness, adherence and internal stresses 178
  3.5.1. Hardness 178
    3.5.1.1. The indentation size effect 179
    3.5.1.2. Hardness tests for coated materials 180
  3.5.2. Coating adhesion 190
    3.5.2.1. Methods for adherence testing 192
  3.5.3. Residual stresses in coatings 201
    3.5.3.1. Origin of internal stresses. 201
    3.5.3.2. Determining residual stresses upon X-ray diffraction 202
    3.5.3.3. Determining internal stresses by radius of curvature measurements (Stoney's method) 202

Bibliography 205

Index 223

Foreword

Tribology is an old science, yet one that remains under active development. Friction between solids was first measured as early as the Renaissance by Leonardo da Vinci who discovered (and recorded in his notebooks) some extraordinary results. Leonardo's laws were subsequently rediscovered by the Frenchman Amontons in the 18th century. However, another 150 years elapsed before they were completely understood thanks to the efforts of the English school led by David Tabor in Cambridge.

Since the early 20th century, we have learnt to reduce friction wear in engines by combining oils with an increasing range of smart additives such as surfactants and polymers. We now have a fair understanding of the adhesion and friction of a tyre on the road, even in wet conditions. We are also able to manufacture rather efficient brakes, albeit by unsophisticated techniques at times (such as with the braking of a high-speed TGV on its tracks using a sand discharge in front of the wheels).

However, new questions are emerging; for example, we poorly understand such phenomena as the onset of earthquakes. To probe surfaces in the laboratory, we have access to highly innovative instrumentation such as the force microscope. We also have powerful simulation tools. Ultimately, from the practical point of view, a whole range of new materials is now available allowing us to reduce friction while at the same time minimizing wear.

It is essential that this know-how be made available to students and practicing engineers alike. In this regard, I find the present book particularly well suited to the needs of both graduate students and professional workers in the field. I therefore wish it every success!

P.-G. De Gennes

Preface

Derived from the Greek word tribos, meaning friction, the word tribology was first used in 1966 in Great Britain to describe the scientific and technical domains focused on the study of friction, wear and lubrication.

Tribology addresses such questions as: what is the best way to reduce wear and control friction? What materials should be used? What lubricant should be chosen to protect a motor or manufacture a particular component? What surface treatment should be applied to improve the wear resistance and reliability of a mechanical system?

Despite their apparent simplicity, the problems of tribology are in fact very complex. They involve the bulk properties of materials as well as their microscopic surface characteristics and their interaction with the surrounding environment.

By providing solutions in fields as diverse as car manufacturing and medical prosthesis, tribology can have a significant economic and ecological impact. For instance, if we can reduce friction in a motor, we reduce energy consumption and limit pollution. If we can reduce the wear in a cutting tool, we both increase productivity and improve the quality of manufactured products. If we can limit ball bearing and gearbox wear, then we can increase the life-span and improve the reliability of many different mechanical systems. If we can reduce friction in an artificial hip, then we can avoid production of wear-particles that can induce inflammation and cause serious complications such as osteolysis or even loosening of the joint itself.

Because tribological phenomena are by nature complex, solving them requires a multidisciplinary approach combining techniques derived from mechanics, solid-state physics and surface chemistry. This sometimes makes it seem like a poorly defined subject, and raises a number of challenges for effective teaching. The purpose of this book is therefore to make tribology comprehensively accessible, by illustrating its principles and its applications through a variety of case studies taken from the scientific as well as the industrial domains. In both its content and structure, *Materials and Surface Engineering in Tribology* is designed to provide a clear, synthetic overview of the field, and therefore serve as a reference book suitable for students, researchers and engineers alike.

Materials and Surface Engineering in Tribology is divided into three chapters:

Chapter 1 introduces the notion of a surface in tribology where a solid surface is described from the topographical, structural, mechanical and energetic points of view. It also describes the principal techniques used to characterize and analyze surfaces.

Chapter 2 discusses what may be called tribology proper by introducing and describing the concepts of adhesion, friction, wear and lubrication.

Chapter 3 focuses on the materials used in tribology. We introduce the major classes of materials used, either in their bulk states or as coatings, including both hard-facing protective layers and other coatings used for decorative purposes.

Acknowledgements

I wish to express my thanks to (in alphabetical order) Patrice Berçot, Lamine Boubakar, Bernard Cretin, Patrick Delobelle, John Dudley, Joseph Gavoille, Jan Lintymer, Hamid Makich, Nicolas Martin, Christine Millot, Jean-François Pierson, Claude Roques-Carmes and Hassan Zahouani for their contributions as proofreaders, for helping with the preparation of figures or for providing resource materials.

Finally, I wish to thank the authors and publishers who granted me permission to reproduce some figures from existing publications.

– Authors: Y. Berthier (Elsevier) [BERT 88, BERT 92], H. Pastor (SIRPE) [PAST 87], S. Mischler (MRS) [BIE 00], F. Palmino (FEMTO-ST Dpt LPMO) (Figure 1.14), A.C. Van Popta and J.M. Brett (SPIE) [VANP 04].

– Publishers: EDP Sciences [ADD 06], Elsevier [ABD 06, BUL 06, BUL 90, BUR 87a, BUR 87b, CHI 96, GAV 02a, GAY 01, GROS 01, JON 84, LIM 87, LIN 03, LIN 04, NEV 99, PIL 06, PIV 94, QUI 02, TAK 87, VOE 96], CNRS Editions/Eyrolles [GEO 00], Société Française du Vide (SFV) [TER 96], Springer [TAK 94b], the American Chemical Society [MART 95] and Wiley [BRU 03].

Chapter 1

Surfaces

1.1. Introduction

The surface of a solid delimits its volume and defines the region where interactions with its environment occur.

When considering the structure of a crystalline material, the surface corresponds to a discontinuity in the periodic arrangement of atoms. The number of nearest neighbors (8 for a body-centered cubic lattice, 12 for a hexagonal close-packed or face-centered cubic lattice) is reduced for surface atoms, so that their vibrational states, inter-atomic separations and associated electronic states are very different from those of atoms within the solid's interior [BOI 87] (see Figure 1.1).

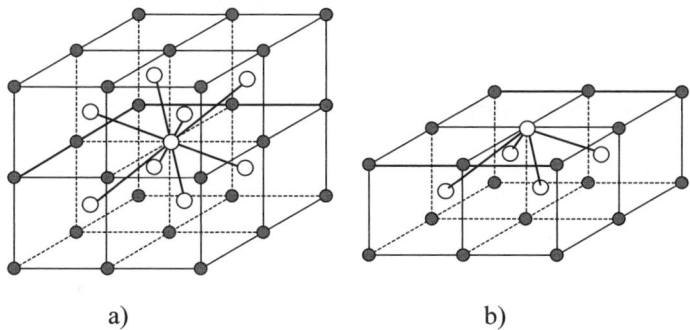

Figure 1.1. *Centered cubic lattice showing the number of nearest neighbors for an atom: a) within a solid; b) on its surface*

This difference between surface atoms and volume (or bulk) atoms allows us to introduce the notions of an ideal surface and a realistic surface. An ideal surface is defined only in terms of the topmost atomic layers spanning a dimension of only several nanometers, whereas a realistic surface describes a region that extends further below the outer surface, down to depths of several microns to several tens of microns. The mechanical, physicochemical and structural properties of this region differ noticeably from those of the material's volume as well as those of the ideal surface [PERR 87].

Because surface atoms possess a lower number of nearest neighbors, they are involved in a smaller number of bonds and thus experience an asymmetric force field. Indeed, these atoms interact only with other surface atoms and atoms situated within the solid's interior. This results in a certain number of dangling bonds, directed towards the exterior of the solid, which allow the solid to interact with its environment through the establishment of bonds aiming to re-establish the surface atoms' equilibrium.

The number of dangling bonds and the nature of their interactions with atoms or molecules in the environment depend on the surface atom coordination and therefore its crystallographic orientation. Each crystallographic plane (each grain in a polycrystalline material) will therefore possess a specific reactivity and physicochemical and mechanical properties different from those of other planes. Passivation and gas adsorption kinetics, speed of corrosion, hardness, Young's modulus, surface energy and electron work function are all amongst the properties that depend on the surface in this way [BERA 83, OUD 73].

1.2. The surface state

A surface can be characterized by its mechanical, physicochemical, topographic or structural properties. The combination of these characteristics defines what we refer to as the surface state.

1.2.1. Structural state of a surface

During shaping or machining processes, the contact between a surface and a machining tool considerably modifies the crystalline structure of the superficial surface layers through mechanical and thermal stresses.

Figure 1.2 shows the cross-section of a surface presenting different zones or layers:

(i) The first zone is a contamination zone consisting of a layer of adsorbed gases such as water vapor, hydrocarbons and other atmospheric pollutants. This surface layer extends to a depth of a few nanometers.

(ii) The second zone is made up of products arising from the interaction with the environment and generally consists of oxides whose composition depends both on the base metal and the environment.

(iii) The third zone corresponds to a material structure that has been significantly "work-hardened" and where the crystalline matrix is essentially destroyed. This layer, often referred to as the Beilby layer, extends to a depth of about one micron.

(iv) The fourth zone is one which has been mechanically deformed from the accumulation of residual stresses. Its thickness ranges from several microns to several tens of microns.

(v) The fifth zone corresponds to the unmodified structure of the original material.

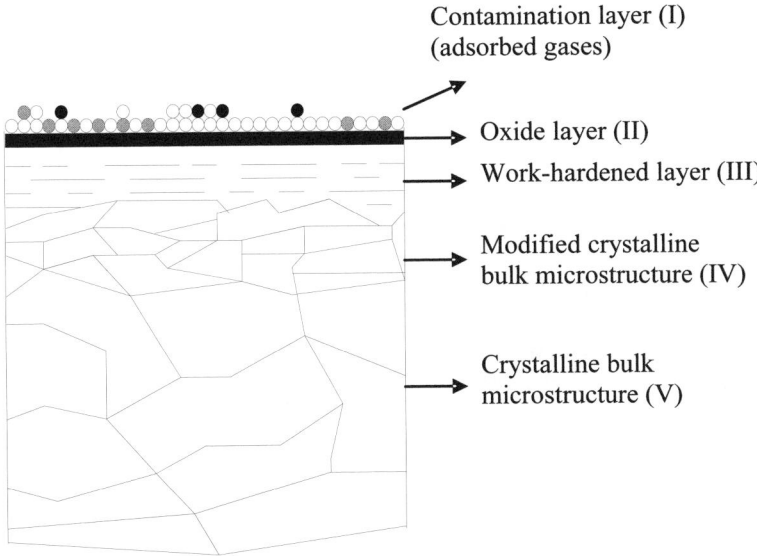

Figure 1.2. *Visual representation of the different layers making up the surface of a material*

1.2.2. *Topographic state of a surface*

It is important to distinguish the topographic state of the surface on the atomic scale from its geometric state on the micrometer scale.

1.2.2.1. *Atomic-scale topographic state [DES 87]*

On the atomic scale, the surface appears as a series of steps/ledges and atomic planes containing numerous point defects. This description of the surface is known as the *Terrace Ledge Kink* (TLK) model illustrated in Figure 1.3. Moreover, depending on their orientation, the crystallographic planes can appear more or less smooth and therefore present varying degrees of roughness (see Figure 1.4).

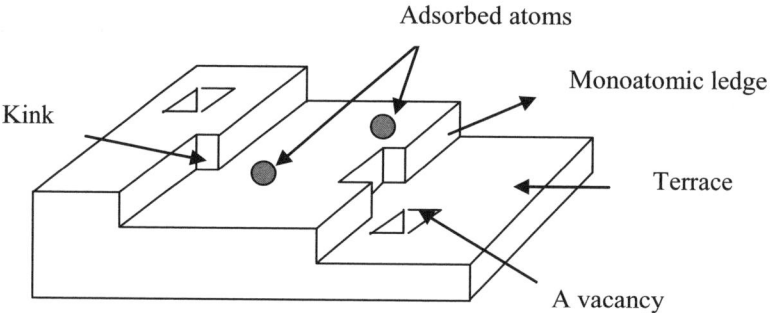

Figure 1.3. *Visual representation of atomic-scale defects on the surface of a solid*

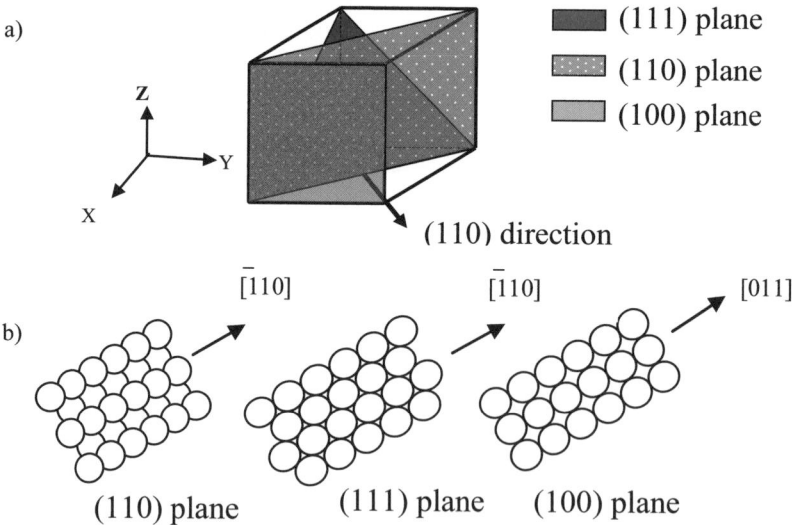

Figure 1.4. *a) Visual representation of several crystallographic planes of a face-centered cubic lattice); b) the roughness of these planes differs*

1.2.2.2. *Micrometer-scale topographic state* *[MIC 89, THOMASTR 99, THOMAST 05]*

Surface machining processes such as turning, milling, polishing or sintering inevitably induce some degree of superficial roughness (see Figure 1.5a) which can be described by a function z(x,y) defining the topographic defects at a point (x,y) as shown in Figures 1.5b and 1.5c.

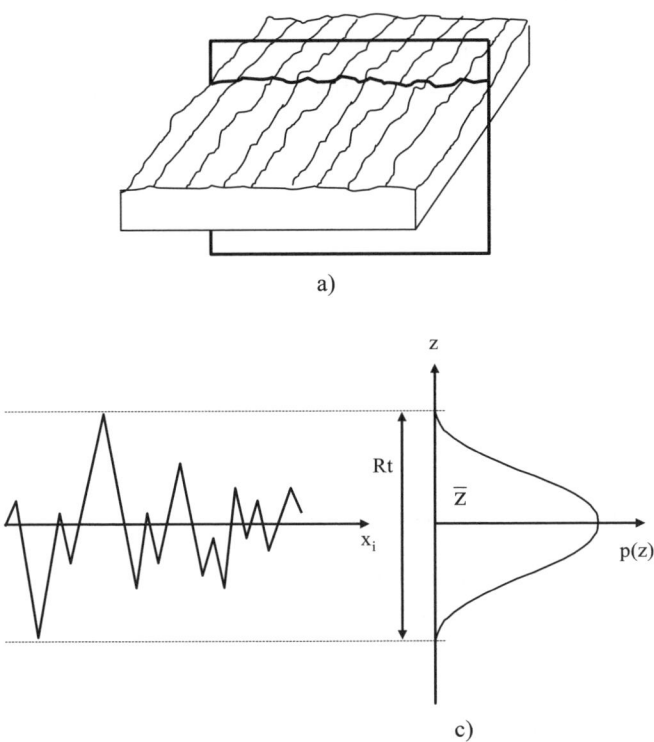

Figure 1.5. *a) 3D analysis of a surface and cross-section of a profile of that surface; b) roughness profile of cross-section (a); c) p(z) is the function of height distribution, ($\bar{z}$) the mean height and (Rt) the total roughness height*

In order to describe the height distribution, we consider the function z(x,y) as a random value, introducing the probability density function p(z) as follows:

$$p(z)dz = \text{prob}\{z \leq z(x,y) \leq z+dz\} \qquad [1.1]$$

where

$$\int_{-\infty}^{+\infty} p(z)dz = 1 \text{ for } p(z) \geq 0 \qquad [1.2]$$

such that p(z)dz represents the probability that the height at a coordinate point (x,y) is between z and z+dz.

The asymmetry parameter (Sk: *Skewness*) and peakedness parameter (Ek: *Kurtosis*) of the height distribution relative to a Gaussian[1] distribution are given in terms of centered moments (m) of order 2, 3 and 4 by the following expressions:

$$Sk = \frac{m^{(3)}}{\left(m^{(2)}\right)^{\frac{3}{2}}} \quad \text{and} \quad Ek = \frac{m^{(4)}}{\left(m^{(2)}\right)^{2}} \qquad [1.3]$$

The nth-order centered moment of p(z) is defined by:

$$m^{(n)} = \int_{z_{min}}^{z_{max}} (z - \bar{z})^n p(z) dz \qquad [1.4]$$

with Sk = 0 and Ek = 3 for a Gaussian distribution.

Skewness characterizes the asymmetry of p(z) relative to a Gaussian distribution. A surface profile where the total area of holes is less than the total area of bumps will give a positive Sk (and a peak-type profile) whereas in the opposite case (with a valley-type profile) Sk will be negative (see Figure 1.6).

Kurtosis characterizes the peakedness of p(z) relative to a Gaussian distribution. For Ek less than 3, the total number of points in the neighborhood of the mean line is greater than that of a normal distribution (flattened profile). Conversely, when Ek is greater than 3, the majority of points are far from the mean curve, which results in a highly peaked profile (see Figure 1.7).

---

1 A random variable x will follow a Gaussian (or normal) distribution if its probability density is expressed in terms of mean m and variance σ^2 as follows:

$$f(x) = \frac{1}{\sigma\sqrt{2\pi}} e^{-\frac{(x-m)^2}{2\sigma^2}}$$

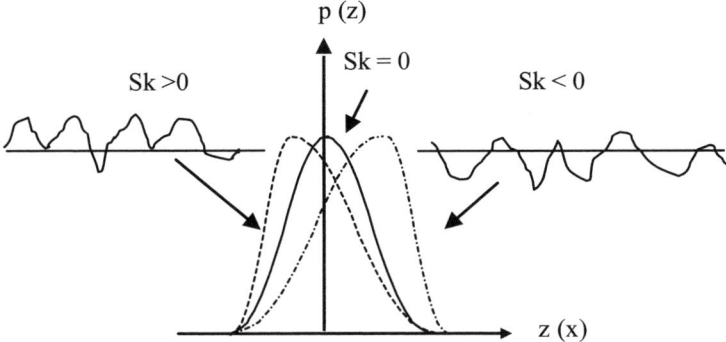

Figure 1.6. *Asymmetry coefficient (Sk) of a roughness profile relative to a Gaussian*

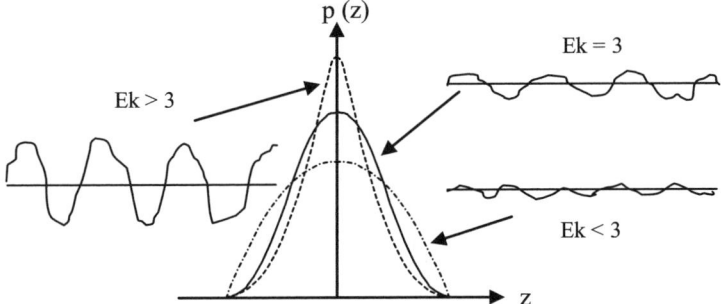

Figure 1.7. *Peakedness coefficient (Ek) of a roughness profile relative to a Gaussian*

We can also define a cumulative probability density function by:

$$P(z) = \int_{z_{min}}^{z_0} p(z)dz \qquad [1.5]$$

where P(z) is the probability for the variable z to be less than or equal to z_0. The quantity $1 - P(z_0)$ represents the fraction of the load-bearing surface of a profile cut at a height of z_0. This corresponds to the bearing ratio at z_0, i.e. the probability that z is greater than or equal to z_0.

A plot of $1 - P(z)$ (expressed as a percentage) shows the evolution of the surface (or profile) bearing ratio as a function of z, and is known as the *bearing area curve* or the *Abbott* or *Firestone curve* (see Figure 1.8).

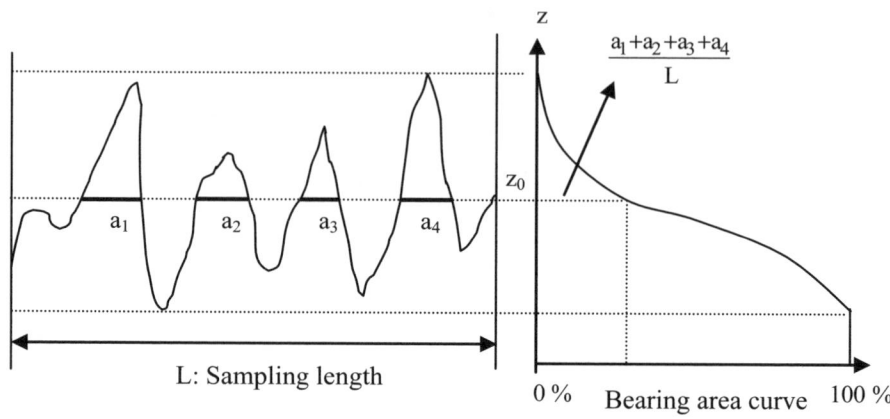

Figure 1.8. *Surface bearing area curve (or Abbott curve)*

1.2.2.2.1. Normalized height distribution parameters

These parameters are given relative to a mean profile line that is defined as the least-squares line:

– Rt: total roughness depth. This corresponds to the difference between the maximum and minimum height in the interval considered.

$$Rt = z_{max} - z_{min} \quad [1.6]$$

– Rp: mean roughness depth. This corresponds to the mean of the distribution z(x) for a given profile of length L:

$$Rp = \frac{1}{L}\int_0^L z(x)dx \quad [1.7]$$

– Ra: mean arithmetic deviation. This is defined by:

$$Ra = \frac{1}{L}\int_0^L |z(x)|dx \quad [1.8]$$

– Rq (or RMS): the root mean square deviation. This is defined by:

$$Rq = \sqrt{\frac{1}{L}\int_0^L z(x)^2 dx} \quad [1.9]$$

1.2.2.2.2. Frequency parameters

In order to analyze periodic surface defects and to quantify the degree of correlation between two given points, we use spectral analysis. In this respect, the basic tool is the Fourier transform of the function z(x):

$$T(s) = \int_{-\infty}^{+\infty} z(x) e^{-2i\pi sx} dx \qquad [1.10]$$

The energy spectral density G(s), defined as the square modulus of the Fourier transform of z(x), can be used to analyze the periodicity of the microgeometric surface state:

$$G(s) = |T(s)|^2 \qquad [1.11]$$

The autocorrelation function $C(\mu)$ is also used to analyze the anisotropy of the state of the surface, and to establish the degree of correlation between two coordinates separated by a distance μ:

$$C(\mu) = \lim_{L \to +\infty} \frac{1}{L} \int_0^L z(x)z(x+\mu)dx \qquad [1.12]$$

where $C(\mu)$ gives the probability of finding two points, at the same height, separated by a distance μ. $C(\mu)$ equals 1 when $\mu = 0$ and decreases pseudo-exponentially, tending asymptotically to zero for large μ. If the surface contains a periodic defect of period λ (induced, for example, by a machining process), the curve $C(\mu)$ will show a peak each time μ is a multiple of λ.

1.2.2.3. *Experimental techniques*

1.2.2.3.1. Stylus profilometers

Stylus profilometers (or 3D profilometers) are devices equipped with a mechanical stylus consisting of a diamond tip whose radius of curvature is generally between 1 and 2 µm. The surface to be analyzed is moved under the stylus by two stepping motors that allow orthogonal movements with micron-level steps. The vertical movements of the stylus follow the topographic defects of the surface and are analyzed using a sensor that generates an electrical signal which is in turn digitized and processed by computer (see Figure 1.9). Figure 1.10 shows an example of 3D topographic measurement of a surface.

10 Materials and Surface Engineering in Tribology

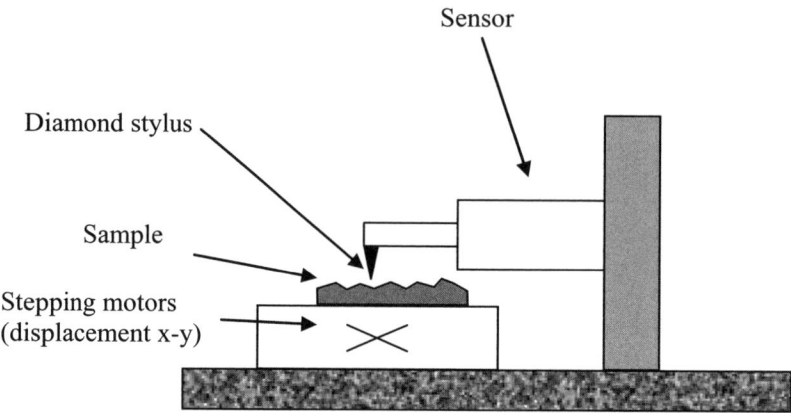

Figure 1.9. *3D stylus profilometer*

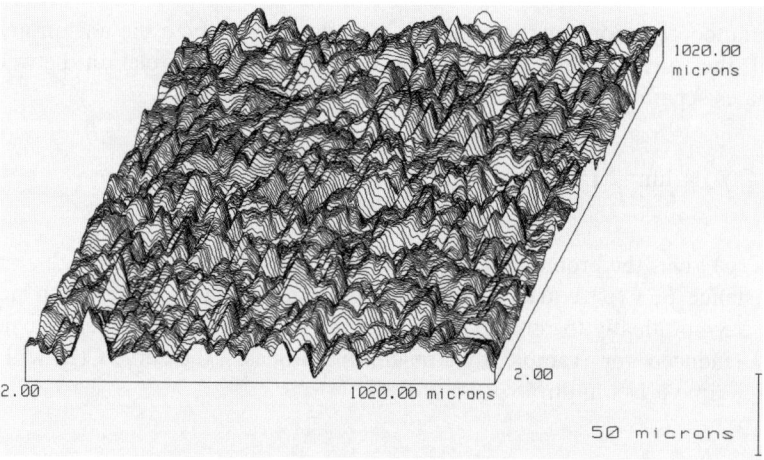

Figure 1.10. *Sandblasted aluminum surface roughness parameters of the surface:*
Ra = 3.1 µm, Rt = 40 µm, Sk = 0.03, Ek = 4.36

1.2.2.3.2. Optical systems [CHENF 00, GAS 95]

Measuring the state of a surface through optical (non-contact) methods uses a laser beam or white light beam to scan the sample surface. In certain cases, it is possible to use "full field" methods employing cameras to avoid the need for scanning.

The confocal microscope

The principle of the laser scanning confocal microscope is shown in Figure 1.11. Small diaphragms (of a few microns in diameter) are placed in front of the laser source and just before the detector, allowing the size of the light source to be reduced to a point. The two diaphragms are placed in conjugate focal planes such that a single lens can focus both incident and reflected beams.

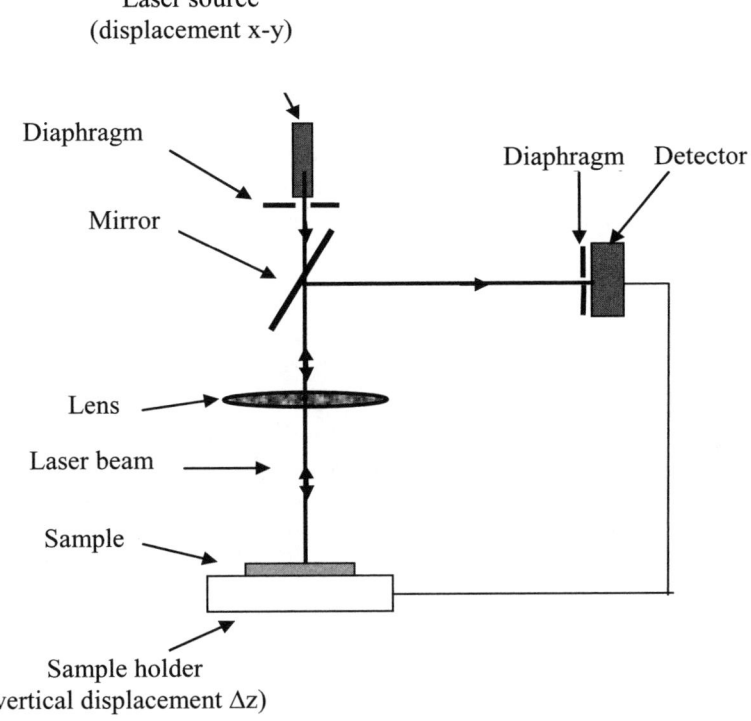

Figure 1.11. *Principle of the confocal microscope*

The laser scans the sample surface point-by-point along a line with a step x, repeated along parallel lines separated by a step y. The beam is reflected from the surface and then focused on the detector. This allows data collection of the 2D topography of the surface xy at a given height z as determined by the focal plane of the laser beam. The height of the sample is then moved by a step Δz and a new phase of data collection is carried out. The 3D representation of the surface is

12 Materials and Surface Engineering in Tribology

obtained through the combination of the data extracted from the different images obtained at different heights z, with vertical and lateral resolutions of the order of 200 nm.

The interferometric microscope

The interferometric microscope consists of an optical microscope coupled to a two-beam optical interferometer of the *Michelson*, *Mirau* or *Linnik* type. In practice, an optical microscope can be converted to an interferometric microscope by replacing the standard objective with an interferometric objective, and by the addition of a high-resolution vertical translation stage.

The device operates as follows: a CDD camera records and digitizes a series of interferograms formed by the combination of a reference beam (reflected from the internal mirror of the interferometric objective) and the beam reflected from the surface of the sample. The interferograms recorded at different heights are processed by specialized software able to reconstruct the surface topography (see Figure 1.12).

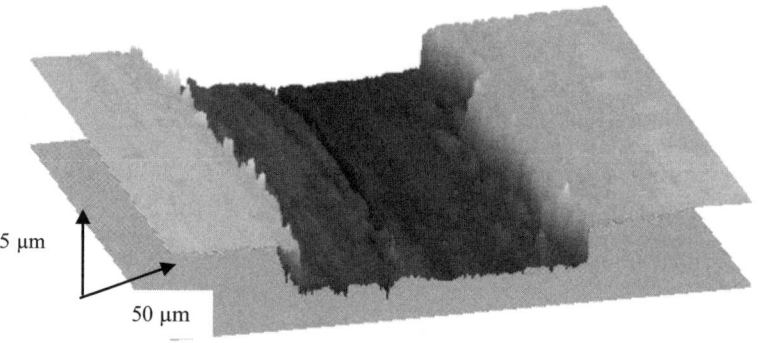

Figure 1.12. *Impression of a spherical aluminum oxide indenter on a copper surface (topographic mapping achieved with an interferometric microscope)*

Inferometric microscopes are easy to set up and have rapid data acquisition times. It takes only a few seconds to map a surface with nanometer-scale resolution.

1.2.2.3.3. Local probe microscopes

Local probe microscopes (or near-field microscopes) are very high resolution instruments for surface characterization that exhibit dramatically improved performance compared with stylus or optically based profilometry. Local probe microscopes exploit very short-range interactions between a fine probe and the sample surface to yield ultra-high resolution, ranging from a micron to a fraction of a nanometer. They systematically use a scanning technique, with the image obtained resulting from the complete scanning of the surface, which can lead to relatively long acquisition times. The resolution of these images essentially depends on two factors: the size of the probe and the way in which the probe–surface interaction varies as a function of the distance to the surface.

Scanning tunneling microscope (STM)

STMs (see Figure 1.13) enable the study of topography and the local properties of metallic and semi-conductor surfaces with angström-level resolution [STI 04]. The principle of operation of the STM is based on measuring the electric current that arises due to the tunneling effect between a fine probe and the sample surface when these are separated by a few angströms (0.3–1 nm), and subject to a potential difference of a few tens of millivolts (10 mV to 1 V). The probes used are generally made of tungsten or of platinum-iridium and have radii of curvature of a few nanometers.

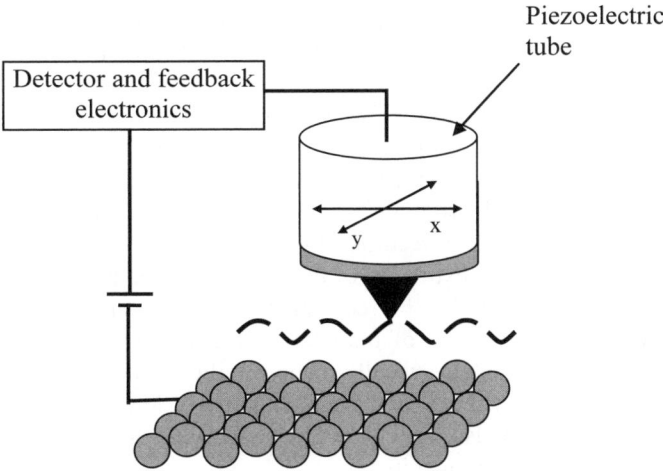

Figure 1.13. *Principle of the scanning tunneling microscope*

As the probe scans the surface, an electronic control system measures the tunneling current and moves the probe tip away from or towards the surface to ensure a constant current intensity (see Figure 1.13). By recording the variations of the distance between the probe and the surface as a function of the coordinates of the scanned points, an atomic-scale 3D representation of the topographic surface is obtained (see Figure 1.14).

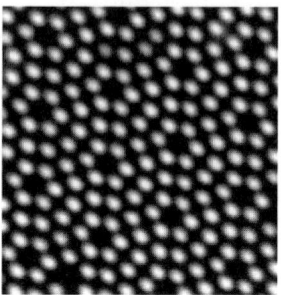

Figure 1.14. *Silicon surface Si (111): the silicon atoms appear clearly on the picture (STM image courtesy of Frank Palmino (FEMTO-ST, LPMO Dpt))*

Atomic force microscope (AFM)

One of the main limitations of the STM is that it cannot be used to analyze non-conducting materials. To overcome this, other techniques such as the atomic force microscope (AFM) have been developed [BIN 86, BIN 87, FRE 04].

This microscope is sensitive to forces resulting from the interactions (see section 2.5.1) between the surface and a fine probe fixed to the tip of a coil spring plate or cantilever with a low spring constant (0.1–10 N m^{-1}). When subjected to these forces (with intensity ranges of 10^{-13}–10^{-6} N) the cantilever is subject to a deflection of a few tens of nanometers which is recorded by measuring the variation of the laser beam position reflected by the extremity of the cantilever (see Figure 1.15). The 3D image of the surface can therefore be reconstructed after computer analysis of the recorded deflections as a function of the coordinates of the points examined (see Figure 1.16). The probes used are usually made of silicon nitride (Si$_3$N$_4$) and have radii of curvature ranging from a few nanometers to a few tens of nanometers. Atomic-scale resolution can also be achieved with this type of microscope, using an ultrafine probe with a single atom at its tip.

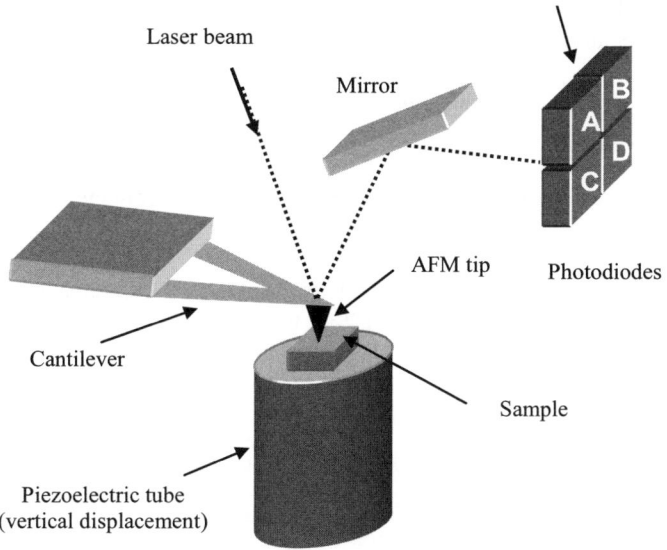

Figure 1.15. *Atomic force microscope*

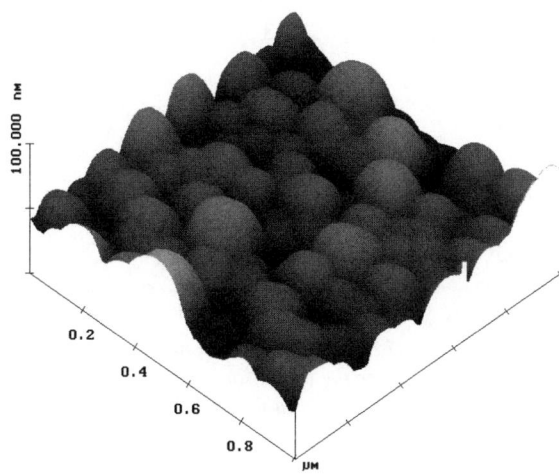

Figure 1.16. *Atomic force microscope image of a copper coating deposited on silicon by reactive magnetron sputtering*

16 Materials and Surface Engineering in Tribology

When in contact mode, the tip of the atomic force microscope remains in contact with the surface and scans its topography. The drawback of this method for soft materials, however, is that it can scratch or deform the sample surface. In such cases, we use the AFM in vibration mode where the probe is maintained a few nanometers from the sample surface and subject to vibratory movement.

In contact mode, it is also possible to measure friction between the probe and the sample. This is referred to as the friction mode and the device used is called a lateral force microscope (LFM) [MAT 87]. As a result of frictional forces, the cantilever is subject to distortions which modify the trajectory of the reflected laser beam (see Figure 1.17).

The use of a four-quadrant photodetector allows the simultaneous detection of the vertical and lateral deviations of the light beam: the former defining the topographic state of the surface and the latter allowing friction measurements (see Figure 1.15).

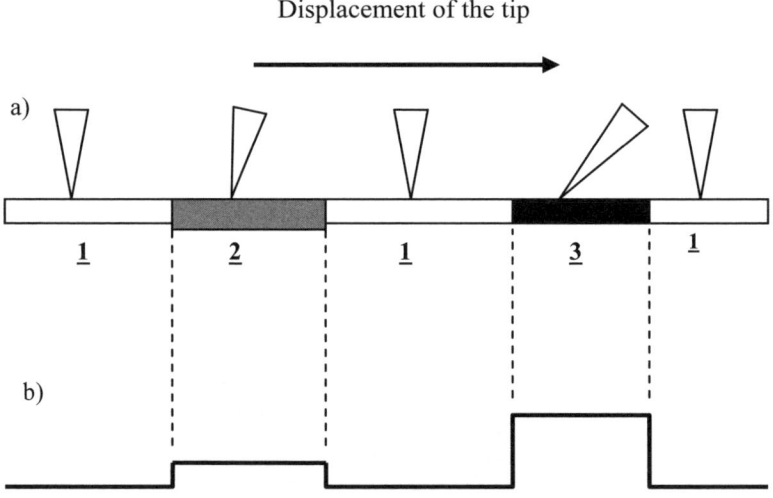

Figure 1.17. *Displacement of the tip of an atomic force microscope on the surface of a sample comprising three areas (1, 2, 3) of a different nature. (a) The higher the friction between tip and surface, the greater the distortion will be for the tip; (b) illustrates the recorded signal and the evolution of lateral forces during the displacement of the tip*

Atomic force microscopes can also detect and map the electrostatic and magnetic forces at the surface of a sample. In order to do this, the AFM is used in vibration mode with the probe placed about 0.5 micron from the surface (far field vibration mode). At that distance, only electric and magnetic forces can be detected. Other forces, such as the Van der Waals, chemical bonds or capillary forces (see section 2.5.1) are only detectable at closer distances (a few tens of nanometers).

Force curves

Force curves show the variations in the interaction force between the tip and the surface within a charge/discharge (or approach/withdrawal) cycle (see Figure 1.18).

When the AFM probe is far from the sample's surface, molecular interaction forces are weak and the cantilever's deflection is near zero (see Figure 1.18a).

When the piezo-electric tube is actioned in order to bring the sample into contact with the probe, two things can occur when the surfaces are sufficiently close together (see Figure 1.15):

1) There is attraction and we observe that the probe jumps towards the surface of amplitude $\delta_1 > 0$. This jump is due to Van der Waals-type interactions and reveals information about their strength (this case is shown in Figure 1.18).

2) There is repulsion between the two surfaces ($\delta_1 < 0$).

If, once contact is established, the sample continues to be displaced towards the probe, we observe a deflection of the cantilever (see Figure 1.18b) that varies linearly with the displacement. Because the cantilever is flexible, it deforms proportionally to the sample displacement, provided the probe does not indent the surface.

If we now consider the withdrawal phase, we observe that the interruption of the probe-sample contact occurs well beyond the position that would correspond to zero force. We note in particular that δ_2, which represents the non-contact jump, is greater than δ_1. δ_2 accounts for the adhesion force between surfaces. With increased adhesion, more energy will be required to break the contact and hence there will be greater deflection of the cantilever δ_2 (see Figure 1.18c).

The adhesion force is simply obtained by multiplying δ_2 by k, the spring constant of the cantilever ($F_{ad} = k\delta_2$).

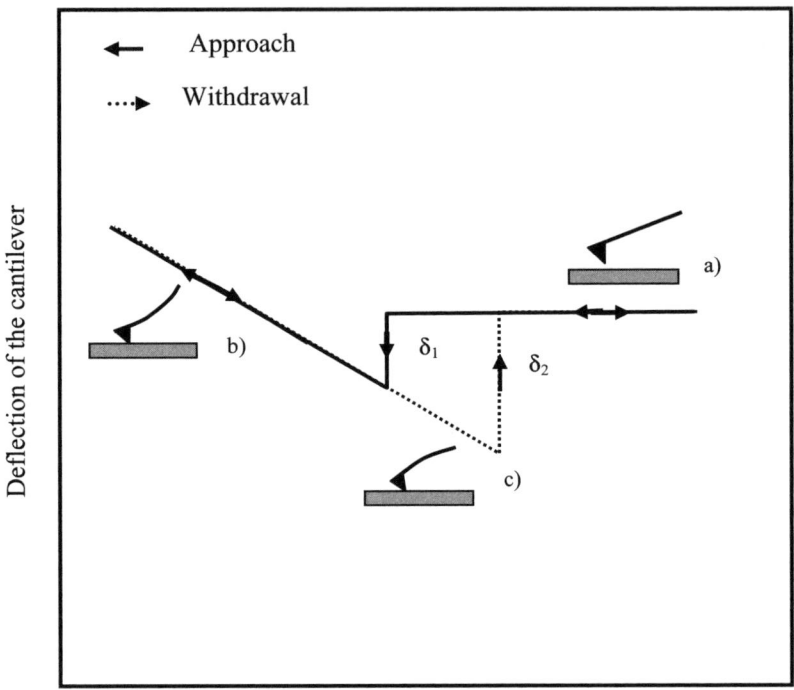

Figure 1.18. *Force curve achieved with an atomic force microscope*

1.2.3. *Surface energy*

A solid's surface energy is a particularly important property in the understanding and prediction of many surface phenomena such as adhesion, friction and bonding. It accounts for the reactivity of materials and is associated with the asymmetric force field to which the surface atoms are subjected. As we have discussed previously, it is this asymmetry that leads to the appearance of dangling bonds that allow the solid's surface to interact with its environment.

The surface energy of a solid is commonly noted γ (J m^{-2}), and corresponds to the energy that must be provided to overcome the cohesion forces of the solid in order to create a new unit area of surface [CARRE 83, COG 00, DAR 97, DEGE 85].

Consider two solids A and B in contact (see Figure 1.19); the energy needed to separate them is called adhesion energy. It is expressed in the fundamental adhesion relationship:

$$W = \gamma_A + \gamma_B - \gamma_{AB} \qquad [1.13]$$

where γ_A and γ_B are the surface energy of A and B, respectively, and γ_{AB} is the interfacial energy.

In the case of a homogenous body, or when the surface energies of A and B are the same, γ_{AB} is zero and the adhesion energy simplifies to:

$$W = \gamma_A + \gamma_B \qquad [1.14]$$

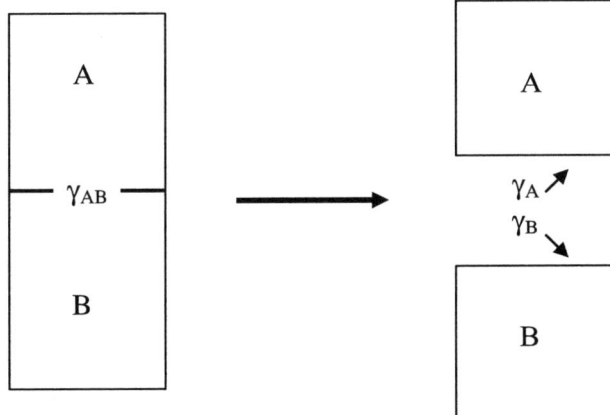

Figure 1.19. *Interfacial energy γ_{AB} characterizes the contact between the two solids A and B; the breaking of the contact yields two new surfaces of surface energy γ_A and γ_B*

A solid's surface energy is therefore directly related to the nature of the bonds between its atoms. These bonds can be chemical (covalent, ionic or metallic) or physical (intermolecular Van der Waals or hydrogen bonds). Covalent materials are characterized by a surface energy between 1000 and 3000 mJ m^{-2}, while ionic crystals have surface energies from 100 to 500 mJ m^{-2}. Molecular crystals have even lower surface energies that are generally less than 100 mJ m^{-2} (see Table 1.1).

Material	Surface energy (mJ m^{-2})
Polytetrafluoroethylene (PTFE)	17
Polyethylene (PE)	32
Polyvinylchloride (PVC)	39
Polymethyl methacrylate (PMMA)	40
Polyamide (PA6)	42
Silicon dioxide	593
Aluminum oxide	1088
Copper (111)	2499
Copper (100)	2892
Tungsten (110)	3320
Tungsten (100)	4680

Table 1.1. *Surface energy values for a few materials*

Surface energy is sensitive to temperature. With increased temperature the network atoms vibrate with increasing amplitude, leading to a decrease in the cohesion energy of the material and a decrease in surface energy.

The adsorption of contaminants on a material's surface, or its oxidation, can also lead to a significant drop in surface energy that can decrease by over an order of magnitude. Mica, for instance, has a surface energy of 5000 mJ m^{-2} in ultra vacuum, while measurements in air yield a value of 300 mJ m^{-2}.

Surface energy is generally calculated from measurements of the contact angle of a liquid with the surface of the solid studied (see Figure 1.20). The contact angle results from the equilibrium between the three interfacial tensions:

γ_{SL} (solid/liquid), γ_{SV} (solid/vapor), γ_{LV} (liquid/vapor)

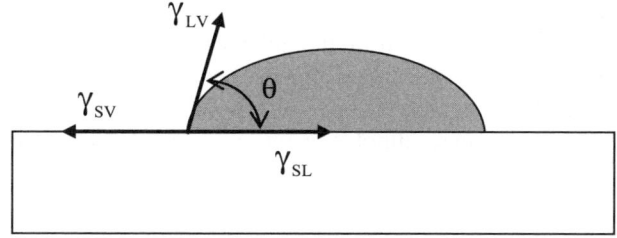

Figure 1.20. *Drop of liquid in equilibrium on the surface of a solid*

The equilibrium of the drop in the presence of the liquid's vapor is represented by Young's equation:

$$\gamma_{SV} = \gamma_{SL} + \gamma_{LV} \cos \theta \qquad [1.15]$$

For the case of liquid–solid contact, equation [1.13] can be written:

$$W = \gamma_{LV} + \gamma_{SV} - \gamma_{SL} \qquad [1.16]$$

Combining equations [1.15] and [1.16] yields:

$$W = \gamma_{LV}(1 + \cos \theta) \qquad [1.17]$$

W represents the adhesion energy between the liquid and the solid. When $\theta = 0$ (perfect wetting of the solid by the liquid), W has its maximum value:

$$W = 2\gamma_{LV} \qquad [1.18]$$

1.2.3.1 *Surface energy measurements*

1.2.3.1.1. The Fowkes [FOW 67], Owens and Wendt [OWE 69] method

The surface energy of a solid γ_{SV} and the surface tension of a liquid γ_{LV} can be written as the sum of two components γ^d and γ^p:

$$\gamma_{LV} = \gamma_{LV}{}^d + \gamma_{LV}{}^p \qquad [1.19]$$

and

$$\gamma_{SV} = \gamma_{SV}{}^d + \gamma_{SV}{}^p \qquad [1.20]$$

where γ^d is the dispersive component of the surface energy and γ^p is the polar component. The former results from the interaction between instantaneous and induced dipoles (London's dispersion forces between non-polar molecules; see section 2.5.1). The latter results from interactions including at least one polar molecule (either two permanent dipoles or one permanent and one induced dipole) (see section 2.5.1).

Table 1.2 gives values for γ_{LV}, γ^p_{LV} and γ^d_{LV} for some liquids.

Liquid	γ_{LV} (mJ m^{-2})	γ^p_{LV} (mJ m^{-2})	γ^d_{LV} (mJ m^{-2})
Water	72.6	51	21.6
Formamide	58.2	18.7	39.5
Diiodomethane	50.8	2.3	48.5
Ethylene glycol	48.3	19	29.3
α-bromonaphtalene	44.6	0	44.6
Tricresylphosphate	40.9	1.7	39.2
Dimethyl formamide	37.5	5	32.5
Octane	21.3	0	21.3
Hexane	18.4	0	18.4

Table 1.2. *Surface tension values (γ_{LV}) for a few liquids; γ^p_{LV} and γ^d_{LV} represent the dispersive and polar components*

Interfacial solid-liquid energy can be expressed as either of the two following equations:

$$\gamma^d_{SL} = \left(\sqrt{\gamma^d_{SV}} - \sqrt{\gamma^d_{LV}}\right)^2 \quad [1.21]$$

and

$$\gamma^p_{SL} = \left(\sqrt{\gamma^p_{SV}} - \sqrt{\gamma^p_{LV}}\right)^2 \quad [1.22]$$

We can therefore write:

$$\gamma_{SL} = \gamma_{SL}^d + \gamma_{SL}^p = \left(\sqrt{\gamma_{SV}^d} - \sqrt{\gamma_{LV}^d}\right)^2 + \left(\sqrt{\gamma_{SV}^p} - \sqrt{\gamma_{LV}^p}\right)^2 \qquad [1.23]$$

By combining equations [1.16, 1.17, 1.19–1.23], we obtain the following:

$$\frac{\gamma_{LV}(1+\cos\theta)}{2\sqrt{\gamma_{LV}^d}} = \left(\sqrt{\frac{\gamma_{LV}^p}{\gamma_{LV}^d}}\right)\sqrt{\gamma_{SV}^p} + \sqrt{\gamma_{SV}^d} \qquad [1.24]$$

The experimental determination of surface energy γ_{SV} is achieved through the measurement of the contact angle θ of two liquids of a known surface tension γ_{LV} and its two components (one polar, one dispersive).

Equation [1.24] will enable us to write a two-equation system which, when solved, will yield the polar and dispersive components of the surface energy. However, greater precision for γ_{SV} will be achieved if several (five to ten) liquids are used.

In this case, if we plot $\gamma_{LV}(1+\cos\theta)/2\sqrt{\gamma_{LV}^d}$ as a function of $\sqrt{\gamma_{LV}^p/\gamma_{LV}^d}$, we obtain a curve with a slope of $\sqrt{\gamma_{SV}^p}$ and with ordinate-axis interception at $\sqrt{\gamma_{SV}^d}$.

1.2.3.1.2. The Good-Van Oss-Chaudhury method [VANO 88a, VANO 88b]

With this approach, we distinguish the dispersive interactions of the Lifshitz-Van der Waals (or LW) type from acido-basic (or AB) type interactions, the latter being a variant of polar interactions and occurring *via* the exchange/interchange of a pair of electrons (Lewis acid). We write:

$$\gamma = \gamma^{LW} + \gamma^{AB} \qquad [1.25]$$

with:

$$\gamma^{AB} = 2\sqrt{\gamma^+\gamma^-} \qquad [1.26]$$

where γ^+ and γ^- are parameters which show the contribution of the electron-acceptor (Lewis acid) and the electron-donor (Lewis base), respectively.

The following equations have been proposed:

$$\gamma_{SL}^{AB} = 2\left(\sqrt{\gamma_{SV}^+\gamma_{SV}^-} + \sqrt{\gamma_{LV}^-\gamma_{LV}^+} - \sqrt{\gamma_{SV}^+\gamma_{LV}^-} - \sqrt{\gamma_{SV}^-\gamma_{LV}^+}\right) \qquad [1.27]$$

and

$$\gamma_{SL}^{LW} = \left(\sqrt{\gamma_{SV}^{LW}} - \sqrt{\gamma_{LV}^{LW}}\right)^2 \quad [1.28]$$

By combining equations [1.16–1.17, 1.25–1.28], we obtain:

$$\gamma_{LV}(1+\cos\theta) = 2\left(\sqrt{\gamma_{SV}^{LW}\gamma_{LV}^{LW}} + \sqrt{\gamma_{SV}^{+}\gamma_{LV}^{-}} + \sqrt{\gamma_{SV}^{-}\gamma_{LV}^{+}}\right) \quad [1.29]$$

The experimental determination of surface energy γ_{SV} is achieved based on the measurement of the contact angle θ of three liquids of a known superficial tension γ_{LV} as well as its three components $\gamma_{LV}^{LW}, \gamma_{LV}^{+}$ and γ_{LV}^{-}.

Equation [1.29] will allow us to write a three-equation system which, once solved, will enable us to obtain the different components of surface energy.

1.2.4. *Mechanical state of a surface*

The mechanical state of a surface can be characterized using four quantities:

1) hardness, which describes the resistance of materials to plastic deformation;

2) Young's modulus and the elasticity limit (or plastic flow threshold), which characterizes a material's elastic properties;

3) toughness, which accounts for the relative brittleness of a material;

4) residual stresses, which play an important part in the resistance of the material to wear and cracking.

1.2.4.1. *Hardness*

Hardness tests consist of using indenters to make an impression on a sample material under a normal load (F) in order to measure the surface area (S) of the residual impression that is left. The hardness is given by the ratio F/S.

Indenters can be of various shapes: a square-based pyramid is needed for the Vickers test while spherical indenters are used for the Brinell test. For the Rockwell test, we use either conical or spherical indenters and Shore hardness measurements use two types of cones. Table 1.3 presents some values for hardness.

Material	HV (or Shore D) hardness
Aluminum oxide	2000
Silicon carbide	2700
Copper	108
Iron	145
Nickel	210
Aluminum	101
Polypropylene	(70)
High-density polyethylene	(66)
Low-density polyethylene	(51)

Table 1.3. *Some mean hardness values (they can vary according to the purity and microstructure of materials); the Vickers scale is used for metals and ceramics hardness and the Shore D hardness scale is used for polymers (in brackets)*

1.2.4.1.1. Vickers hardness

The indenter used is a diamond square-based pyramid with an angle of 136° between opposite faces as shown in Figure 1.21.

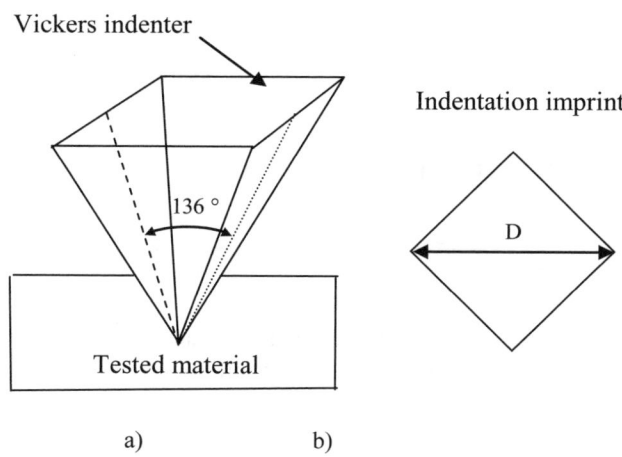

Figure 1.21. *Vickers hardness test: a) square-based pyramidal Vickers indenter; b) residual square indentation of diagonal D*

Vickers hardness is given by the formula:

$$HV = 1.854 \frac{F}{D^2} \qquad [1.30]$$

where F is the applied load (kg) and D is the diagonal of the residual indentation (mm).

1.2.4.1.2. Brinell hardness

Brinell hardness is determined with tungsten carbide spherical indenters of 10, 5, 2.5 or 1 mm in diameter (see Figure 1.22).

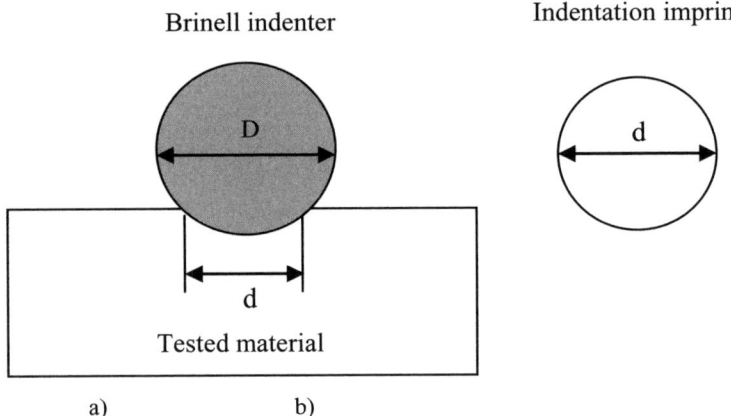

Figure 1.22. *Brinell hardness test: a) Brinell indenter; b) residual impression*

Brinell hardness can be expressed as follows:

$$HB = \frac{0.204 F}{\pi D \left(D - \sqrt{D^2 - d^2} \right)} \qquad [1.31]$$

where F is in newtons and both the diameter of the spherical indenter (D) and the diameter of the residual impression (d) are in millimeters.

1.2.4.1.3. Rockwell hardness

When performing a Rockwell hardness test, it is the depth of the residual indentation impression h, rather than its diameter, which is measured.

We first apply a (minor) preliminary load of 98 N that forces the indenter into the surface to a depth of l. This degree of penetration is not taken into account in the hardness measurement, but is taken as the origin or zero penetration reference point.

We then apply a (major) normalized load to the surface for a few seconds leading to a total penetration t. The major load is then reduced to the minor load and the depth of the residual impression h is measured and given in millimeters.

With the Rockwell B test, we use a steel spherical indenter of diameter 1.58 mm and a normal load of 980 N. With the Rockwell C test, the load is 1470 N and we use a conical diamond indenter with radius of curvature of 0.2 mm and apex angle of 120°.

Both the Rockwell B and Rockwell C hardnesses can be written:

$$HRB = 130 - \frac{h}{0.002} \quad \text{and} \quad HRC = 100 - \frac{h}{0.002} \qquad [1.32]$$

where h is measured in mm. The constant 0.002 has units in millimeters; therefore the Rockwell hardness is a dimensionless number. It simply gives the ratio between the depth of penetration of the indenter and the value of 0.002 mm, relative to an arbitrary constant (130 for HRB hardness and 100 for HRC hardness).

1.2.4.1.4. Shore hardness

The Shore hardness test is specifically suited to polymers such as rubber, thermoplastics or elastomers. The test is conducted with an apparatus called a durometer and employs a calibrated spring that generates a known force on the indenter. The apparatus records the penetration depth of the indenter and shows the corresponding Shore hardness value on a scale graduated from 0 to 100 with these values corresponding to maximum penetration and zero penetration (maximum hardness), respectively.

There are two types of indenters:
1) a truncated cone with an apex angle of 35° (Shore A); and
2) a sharp cone with an apex angle of 30° (Shore D).

1.2.4.2. *Young's modulus*

For an isotropic material, Young's modulus is the constant of proportionality between the stress applied to a rod and its elastic deformation. It is generally obtained from uniaxial tensile tests but, as we will see later, it can also be determined from indentation tests.

28 Materials and Surface Engineering in Tribology

Consider a cylindrical rod of cross-sectional area S_0 and length L_0 to which we apply a force F (see Figure 1.23). The stress $\sigma = F/S_0$ will vary as a function of the strain $\varepsilon = \Delta L/L_0$ following the classic curve shown in Figure 1.24. The stress $\sigma = F/S_0$ and strain $\varepsilon = \Delta L/L_0$ are both normalized relative to the initial dimensions (S_0 and L_0) of the rod. The true stress true strain curve is obtained by considering the true stress $\sigma = F/S$ and the true strain $\varepsilon = \Delta L/L$, where S and L represent the real dimensions of the rod at any instant during the test. The true stress true strain curve deviates from the classic stress–strain curve for increasing strain.

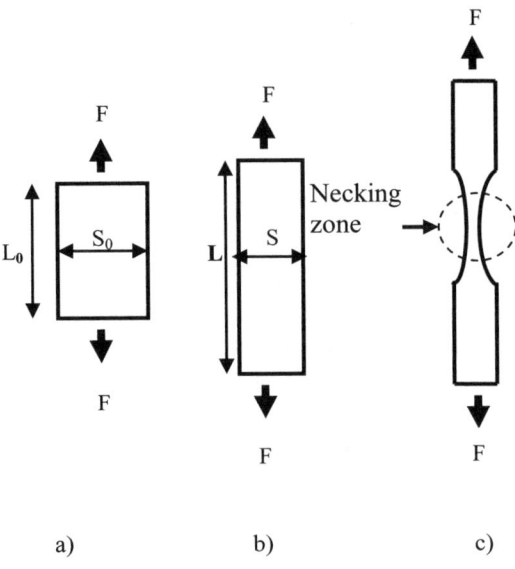

Figure 1.23. *Tensile test: a) at the onset of the test on a cylindrical rod; b) homogenous strain; c) onset of necking*

Figure 1.24 can be divided into three regions. Initially (region OP in the figure), stress and strain are linearly proportional and their relationship can be expressed using Hooke's law:

$$\sigma = E\varepsilon \qquad [1.33]$$

where E is the Young's modulus.

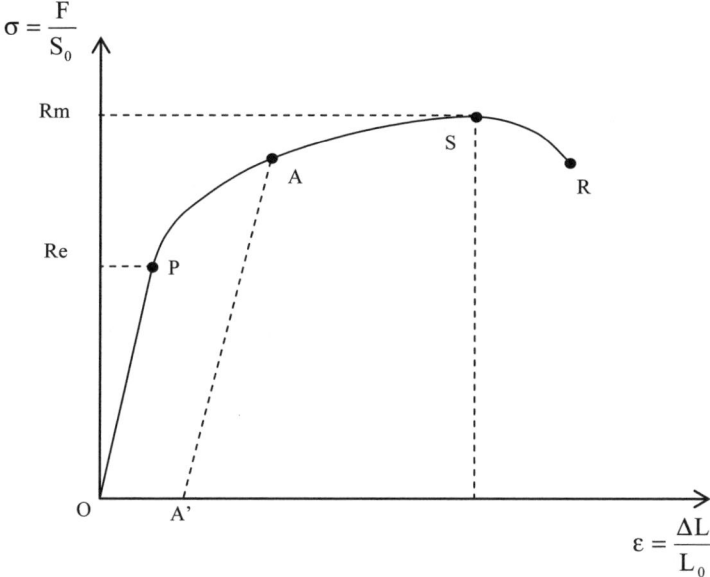

Figure 1.24. *Stress-strain curve*

This linear region is characterized by its reversible nature. If the applied stress is removed, the rod will return to its original dimensions. This is what is known as purely elastic strain. However, when the applied stress reaches a certain threshold value (Re) referred to as the yield strength, deformation of the material becomes irreversible (point P). If the applied stress is removed at the point A, the material will not return to its original dimensions but will show evidence of residual strain (OA').

The region PS of the curve corresponds to irreversible or plastic deformation and, in this region, deformation is uniform along the rod.

The point S corresponds to the onset of necking, the localized narrowing of the rod cross-section. Plastic deformation is then no longer homogenous and the local narrowing of the cross-section (usually in the center of the rod) is accompanied by an increase in the stress, which passes through a maximum before decreasing to its rupture point (R). The value of stress at point S is known as the ultimate tensile strength of the material (Rm). Table 1.4 summarizes the values for E, Re and Rm for some common materials.

Material	E (GPa)	Re (Mpa)	Rm (MPa)
Diamond	1,000	50,000	–
Silicon carbide	420	10,000	–
Aluminum oxide	350	5,000	–
Silicon nitride	260	8,000	–
Nickel	220	70	400
Iron	190	50	200
Copper	124	60	400
Silicon	107	–	–
Silica glass	94	7,200	–
Gold	82	40	220
Silver	76	55	300
Lead & alloys	14	11–55	14–70
Polystyrene	3-3.4	34–70	40–70
Polycarbonate	2.6	–	–
Polypropylene	0.9	19-36	33-36
Very high density polyethylene	0.7	20-30	37

Table 1.4. *Some values for Young's modulus (E), yield strength (Re) and ultimate tensile strength (Rm)*

1.2.4.3. *Nano-indentation*

This technique enables the simultaneous characterization of both the hardness and Young's modulus of the superficial layers of materials for depths ranging from a few nanometers to several microns.

Unlike classic hardness tests which rely on a fixed load, nano-indentation uses a dynamic load (P) which increases linearly with time and which enables the monitoring of the indenter penetration (h) during the phases of loading and unloading.

Nano-indentation tests use a square-based pyramidal Vickers-type indenter as shown in Figure 1.25a or, more commonly, a Berkovich-type indenter (triangular pyramid) as shown in Figure 1.25b.

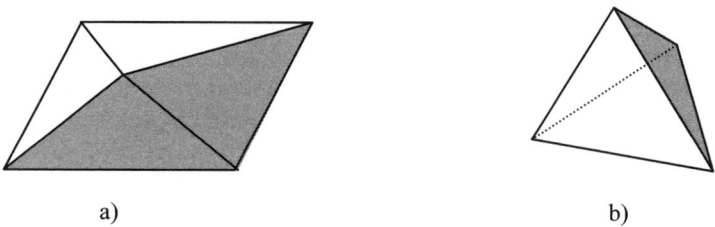

a)   b)

Figure 1.25. *Indenters used in nano-indentation tests:
a) Vickers indenter; b) Berkovich indenter*

Figure 1.26 shows the impression left by the indenter under load and in the absence of a load.

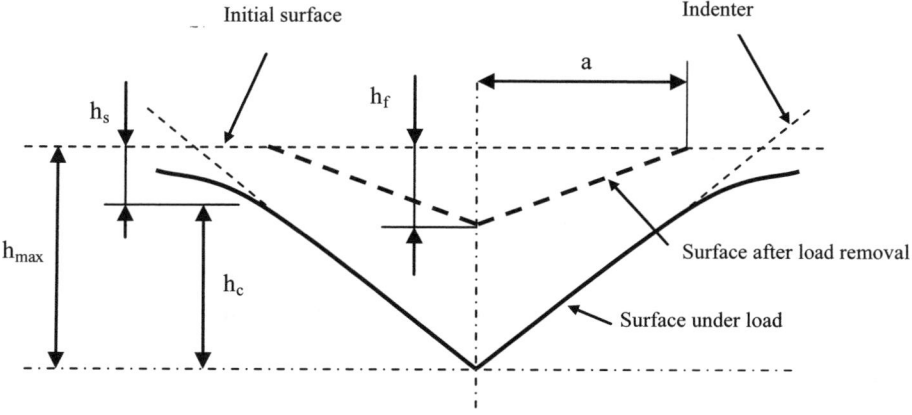

Figure 1.26. *Schematic representation of the cross-section of an indentation
under load and after unloading*

Here h_{max} indicates the indenter displacement at maximum applied load (P_{max}). Experimental tests [OLI 92] carried out on a variety of materials (such as aluminum oxide, glass, sapphire and tungsten) have shown that the load (P) and the displacement (h) can be related by:

$$P = A(h-h_f)^m \qquad [1.34]$$

In this expression, h_f gives the residual indentation depth (in the absence of a load). A and m are constants determined from the fitting of the experimental results obtained for the materials mentioned above. Their values lie within the range 0.0215–0.265 for A and 1.25–1.51 for m.

Table 1.5 gives the values of m for various indenter shapes. For Berkovitch indenters, the value is 1.5.

Indenter shape	m	α
Flat cylindrical indenter	1	1
Paraboloide of revolution	1.5	0.75
Conical indenter	2	0.72

Table 1.5. *Theoretical values for m and α (see equation [1.41]) for three types of axially-symetric indenters [OLI 92]*

During the withdrawal of the indenter, it remains in contact with the surface for a period corresponding to its elastic recovery. This is shown in the linear section of the unload curve of Figure 1.27.

The slope of this line, given by the derivative dP/dh at point (P_{max}, h_{max}), represents the elastic stiffness of the surface and can be expressed as:

$$S = \frac{dP}{dh} = \frac{2}{\sqrt{\pi}} E \sqrt{A_c} \qquad [1.35]$$

where P (N) indicates the load, h (m) the depth of penetration, A_c (m^2) the contact area and E (Pa) the reduced Young's modulus, expressed as:

$$\frac{1}{E} = \left(\frac{1-v_1^2}{E_1} + \frac{1-v_2^2}{E_2} \right) \qquad [1.36]$$

where E_1 and E_2 give the Young's modulus and v_1, v_2 the Poisson coefficients of the indenter and the material's surface, respectively.

Figure 1.27 shows an example of a load/unload curve. The analysis of this curve is based on the model proposed in [OLI 92, OLI 04].

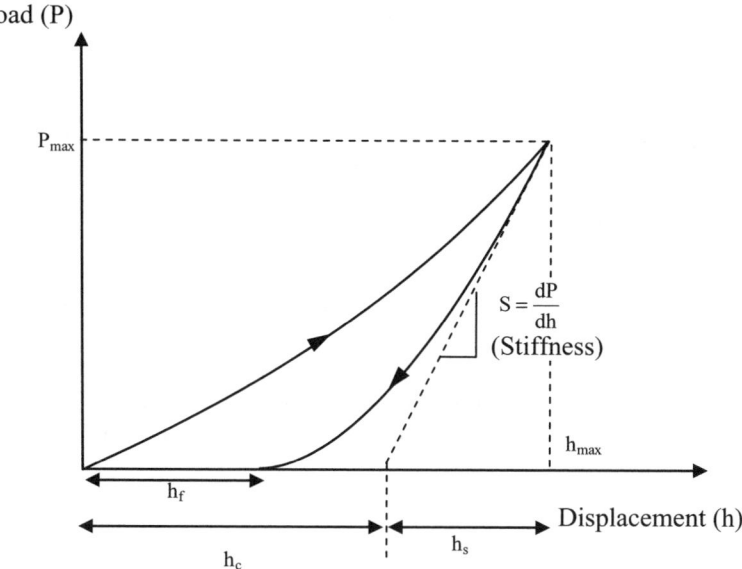

Figure 1.27. *Indentation curve*

In the case of a Berkovitch indenter (the most commonly used in nano-indentation tests), a correction factor ($\beta = 1.05$) is introduced in order to take into account its asymmetric profile:

$$S = \frac{dP}{dh} = \beta \frac{2}{\sqrt{\pi}} E \sqrt{A_c} \qquad [1.37]$$

For this indenter (assuming perfect geometry), the contact area A_c is given by:

$$A_c = 24.5 h_c^2 \qquad [1.38]$$

In reality, the tip of the indenter always exhibits some geometrical imperfection which we include through an additional correction term:

$$A_c = 24.5 h_c^2 + \sum_{i=1}^{8} C_i h_c^{1/(2^i-1)} \qquad [1.39]$$

The coefficients C_i are constants which can be determined using a procedure described in [OLI 92].

Based on Figures 1.26 and 1.27, we can write:

$$h_c = h_{max} - h_s \qquad [1.40]$$

where h_s represents the elastic penetration which can be easily calculated using the expression:

$$h_s = h_{max} - \alpha \frac{P_{max}}{S} \qquad [1.41]$$

where $\alpha = 0.75$ is a geometrical factor for a Berkovich indenter (see Table 1.5).

The determination of S from the unloading curve allows h_s to be calculated from equation [1.41] and thus allows the determination of h_c ($h_c = h_{max} - h_s$). A_c can then be easily calculated and the reduced elasticity modulus E can be obtained (see equation [1.37]):

$$E = \frac{\sqrt{\pi}}{2\beta\sqrt{A_c}} \times S = \frac{\sqrt{\pi}}{2\beta\sqrt{A_c}} \times \frac{dP}{dh} \qquad [1.42]$$

Given the reduced modulus and the mechanical properties of the indenter (E_1 and ν_1), it is then easy to determine the value of the ratio $(1-\nu_2^2)/E_2$ from equation [1.36].

If the precise value of ν_2 is not known, we generally take a value 0.30 for metals and 0.20 for glass and ceramics to calculate the Young's modulus E_2 of the material.

1.2.4.4. *Fracture toughness*

Figure 1.28 shows the different types of fractures which can occur within a brittle material. Brittleness (the opposite of ductility) refers to the ease with which fractures can propagate within a material. Brittleness is quantified through a parameter known as toughness (denoted Kc), which describes the resistance of the material to fracture propagation.

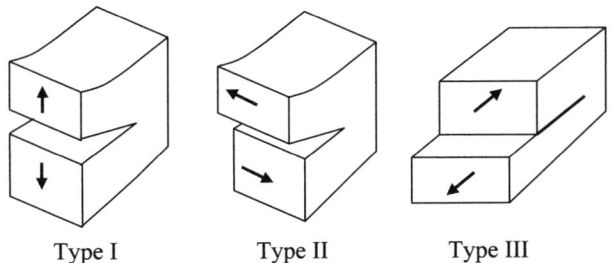

Figures 1.28. *The different types of fractures*

With brittle materials such as glass, minerals or ceramics, the energy of fracture propagation corresponds to the work needed to break the bonds ensuring the cohesion of the solid along a plane. The corresponding energy (W) is then equal to double the surface energy (see equation [1.14]):

$$W = 2\gamma \qquad [1.43]$$

If the energy released during the propagation of the fracture (due to the release of elastic energy) is greater than or equal to that of the creation of a new crack surface (due to the opening of the crack), then the fracture is able to propagate.

In the case of a material of Young's modulus E with a fracture of length 2a (see Figure 1.29) subject to stress σ, the condition for fracture propagation is:

$$\sigma \geq \sqrt{\frac{2E_\gamma}{\pi a}} \qquad [1.44]$$

This can also be written:

$$\sigma\sqrt{\pi a} \geq \sqrt{2E_\gamma} \qquad [1.45]$$

The first term in this inequality $K = \sigma\sqrt{\pi a}$ is the stress intensity factor and the second term $Kc = \sqrt{2E\gamma}$ defines the fracture toughness, i.e. the amount of stress required to propagate a pre-existing crack. Table 1.6 lists some values for fracture toughness.

36 Materials and Surface Engineering in Tribology

Materials	Kc (MPa m$^{-1/2}$)
Aluminum oxide	5
Silicon carbide	3
Copper	200
Iron	160
Nickel	100
Aluminum	250
Polypropylene	3
High-density polyethylene	1
Low-density polyethylene	2.5

Table 1.6. *Some values for fracture toughness*

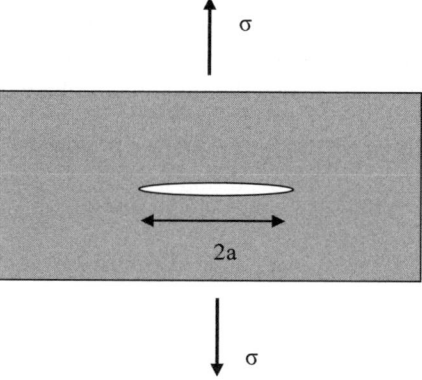

Figure 1.29. *Crack of length 2a within a material subject to stress σ*

A pre-existing crack will propagate, leading to material failure, when K reaches the critical value Kc. Kc is generally determined from impact tests on a single-edge notch bend specimen. However, it can also be determined using Vickers indentation tests. Figure 1.30 shows the shape of an impression obtained on a brittle material. The diagonal of the residual impression (D) allows the calculation of the material's hardness, while the length of the radial fracture (C) allows the calculation of fracture toughness Kc using the following expression [ANS 81]:

$$Kc = 0.016\sqrt{\frac{E}{H}}FC^{-3/2} \qquad [1.46]$$

where F is the applied load and E and H are the Young's modulus and the Vickers hardness for the material, respectively.

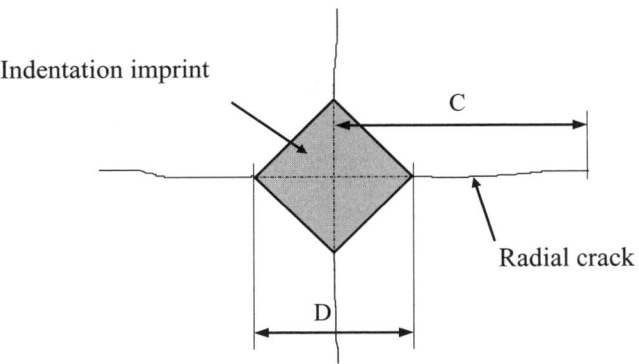

Figure 1.30. *Vickers indentation of a brittle material*

1.2.4.5. *Residual stresses*

Machining processes and different surface production processes induce, on the superficial layers of materials, strains which generate internal "residual stresses". They can generally be classified into two classes: tension or compression stresses. The first class is generally hazardous because they facilitate fracture propagation and can cause breakup of materials and structures; conversely, the second class is beneficial and indeed is often introduced deliberately through surface treatments such as shot peening (see section 3.3.1.6) in order to harden the surface of a material and improve its resistance to wear (see Figure 1.31).

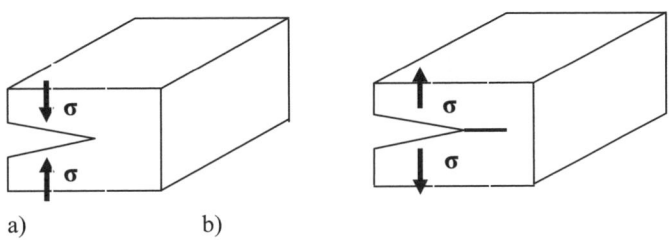

Figure 1.31. *Effects of residual stresses (σ) on fractures: a) compression stress prevents crack propagation; b) strain stress facilitates crack propagation*

Several experimental techniques allow the measurement of residual stress in materials. A first class of so-called "destructive" techniques is based on the cutting-away of some stressed material and the measuring of the resulting deformation in the adjacent material as this is due to the relaxation of the residual stresses. A second class of so-called "non-destructive" techniques is based on the analysis of the modifications of the physical and crystallographic properties of the material induced by residual stresses. Non-destructive techniques include the ultrasound method for example, which uses the influence of residual stress on the variation of the propagation speed of ultrasound waves; the so-called "Barkhausen" technique, based on the interaction between elastic strain and magnetization (displacement of Bloch walls) in ferromagnetic materials; and the method of X-ray diffraction which we will now discuss in more detail.

1.2.4.5.1. *Determination of residual stress by X-ray diffraction* [MAC 86, MAE 88, SPR 80]

Crystalline networks are well-known to diffract X-rays when a family of crystallographic planes of interplanar spacing d_{hkl} (simply denoted d) satisfies the Bragg condition (see Figure 1.32):

$$2d \sin \theta = \lambda \qquad [1.47]$$

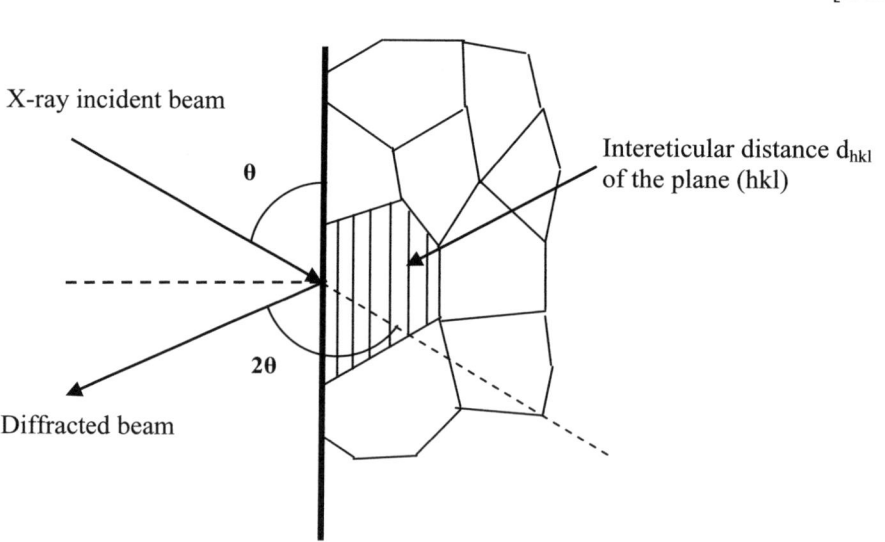

Figure 1.32. *Principle of crystallographic X-ray diffraction*

Differentiating this relation gives:

$$\frac{\Delta d}{d} = -\frac{\Delta \theta}{\tan \theta} \qquad [1.48]$$

If the material is subjected to stress that induces lattice deformation by an amount Δd as shown in Figure 1.33, the relative strain may then be related to the variation in the diffraction angle θ by the equation:

$$\varepsilon = \frac{\Delta d}{d} = -\frac{\Delta \theta}{\tan \theta} \qquad [1.49]$$

We can therefore write:

$$\Delta 2\theta = -2\varepsilon \tan \theta \qquad [1.50]$$

Equations [1.49] and [1.50] show that a variation of d will invariably induce a variation of ε and, consequently, an angular displacement $\Delta 2\theta$ of the diffraction peak, as shown in Figure 1.34.

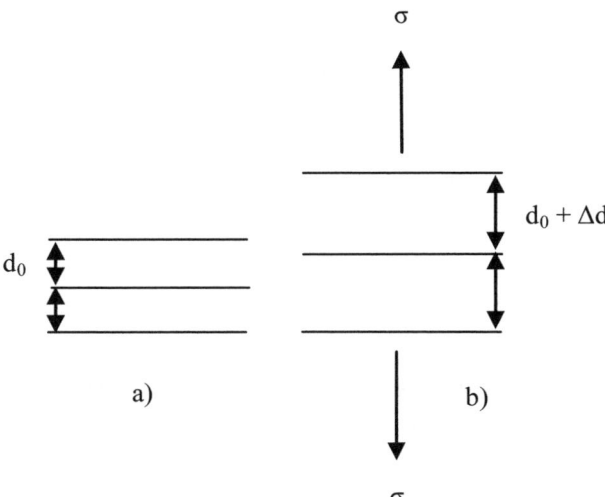

Figure 1.33. *Impact of tractive stress on lattice deformation:*
a) unstrained crystalline network (interplanar spacing d_0);
b) strained crystalline network (interplanar spacing $d_0 + \Delta d$)

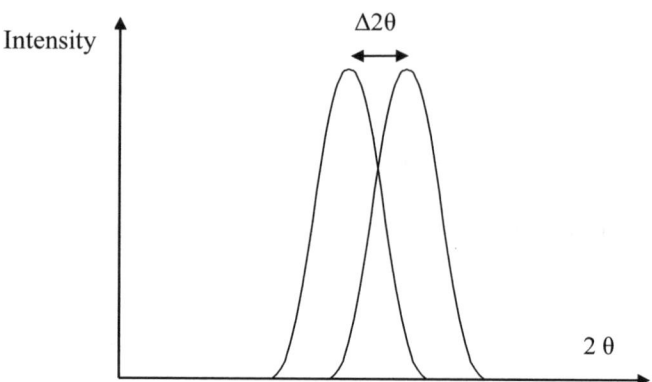

Figure 1.34. *Displacement of the diffraction peak*

The strain ε can be defined by its three components ε_1, ε_2, ε_3 (see Figure 1.35) and will be denoted $\varepsilon_{\varphi\psi}$.

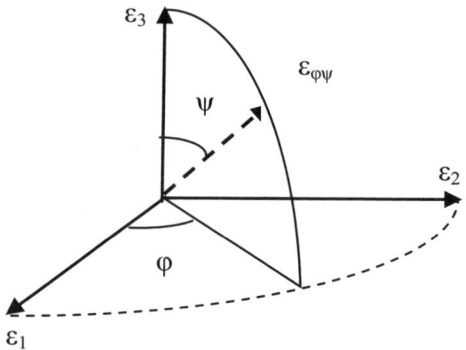

Figure 1.35. *Determination of strain $\varepsilon_{\varphi\psi}$*

If d_0 is the interplanar spacing of the unstrained crystal and d_ψ is the interplanar spacing after deformation, we can write:

$$\varepsilon_{\varphi\psi} = \frac{d_\psi - d_0}{d_0} \quad [1.51]$$

where ψ is the angle between the normal to the sample surface and the normal to the crystallographic planes that diffract the X-rays (see Figure 1.36).

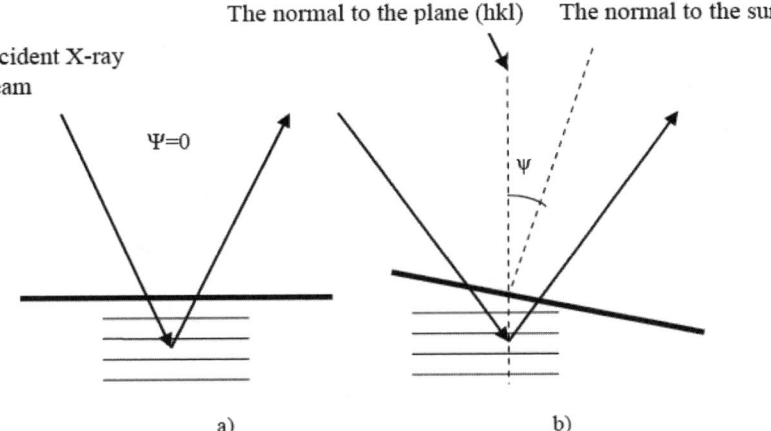

Figure 1.36. *X-ray diffraction by planes (hkl): a) ψ = 0; b) ψ ≠ 0*

When these two planes are parallel we have ψ = 0, and if $d_\perp$ is then the interplanar spacing of the strained group of planes (hkl), the strain is given by:

$$\varepsilon_\perp = \frac{d_\perp - d_0}{d_0} \qquad [1.52]$$

Equations [1.51] and [1.52] then yield:

$$\varepsilon_{\varphi\psi} - \varepsilon_\perp = \frac{d_\psi - d_0}{d_0} - \frac{d_\perp - d_0}{d_0} = \frac{d_\psi - d_\perp}{d_0} \qquad [1.53]$$

If the strain is small, we can write $d_0 = d_\perp$ with negligible error. We can then obtain:

$$\varepsilon_{\varphi\psi} - \varepsilon_\perp = \frac{d_\psi - d_\perp}{d_\perp} \qquad [1.54]$$

If we now use Hooke's law, which allows us to express the strain in terms of the stress, by assuming planar stresses (which is justified given the limited depth analyzed by X-ray diffraction) the strain $\varepsilon_{\varphi\psi}$ can be written:

$$\varepsilon_{\varphi\psi} = \frac{1+v}{E}\sigma_\varphi \sin^2\psi - \frac{v}{E}(\sigma_1+\sigma_2) \qquad [1.55]$$

where σ_1, σ_2 and σ_φ are defined in Figure 1.37.

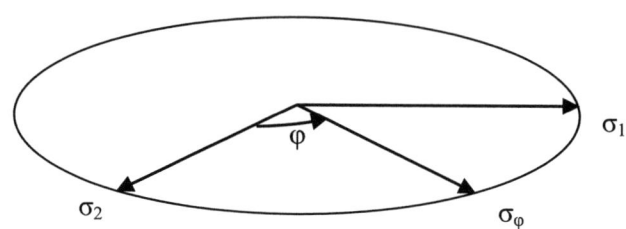

Figure 1.37. *Definition of σ_φ; stress σ_3 is null (state of plane stress)*

If $\psi = 0$, we can use equation [1.55] and the definition of $\varepsilon_\perp$ to write:

$$\varepsilon_\perp = -\frac{v}{E}(\sigma_1+\sigma_2) \qquad [1.56]$$

If we combine equations [1.54], [1.55] and [1.56], we obtain:

$$\sigma_\varphi = \frac{d_\psi - d_\perp}{d_\perp} \frac{E}{1+v} \frac{1}{\sin^2\psi} \qquad [1.57]$$

Equation [1.49] then allows us to write:

$$\varepsilon = \frac{\Delta d}{d} = -\frac{\Delta 2\theta}{2\tan\theta} \qquad [1.58]$$

It is therefore possible to obtain:

$$\frac{\Delta d}{d} = \frac{d_\psi - d_\perp}{d_\perp} = -\frac{2\theta_\psi - 2\theta_\perp}{2\tan\theta_\perp} \qquad [1.59]$$

Equation [1.57] then becomes:

$$\sigma_\varphi = \frac{2\theta_\psi - 2\theta_\perp}{2\tan\theta_\perp} \frac{E}{1+v} \frac{1}{\sin^2\psi} \qquad [1.60]$$

We can write equation [1.60] in the form:

$$\Delta 2\theta = 2\theta_\psi - 2\theta_\perp = 2\sigma_\varphi \tan\theta_\perp \frac{1+\nu}{E} \sin^2\psi \qquad [1.61]$$

Equation [1.61] is known as the "$\sin^2\psi$ method". It gives the variation of $\Delta 2\theta$ as a function of $\sin^2\psi$ in terms of a straight line that passes through the origin with a slope that allows us to calculate the surface strain σ_φ. Experimentally, to obtain σ_φ, $\Delta 2\theta$ will have to be measured for two different angles ψ. However, experience shows that a minimum of five different values of ψ is required to obtain sufficient precision.

The limits of the X-ray diffraction method in determining internal stresses arise directly from the hypotheses necessary to obtain the $\sin^2\psi$ method. The material must be homogenous, continuous and isotropic, and the stresses must be elastic so that the use of Hooke's law is justified. The stresses and strains must also be homogenous throughout the surface analyzed and the stress state needs to be biaxial, implying that the normal stress σ_3 is zero [SPR 80].

When one or more of these requirements are not satisfied, significant deviations are observed in relation to equation [1.61]. This is particularly true in the case of strong anisotropy (as is the case with highly-textured materials) or for large grain polycrystalline materials.

1.2.5. *Chemical composition of a surface*

The chemical composition of a surface can be characterized using a number of different analysis techniques with the same underlying principle. Specifically, an electronic, optical or ionic probe (the primary particle beam) is directed at the surface under study and induces the emission of secondary or backscattered particles (photons, electrons or ions). The characterization of these secondary particles (ion mass, electron velocity, photon wavelength) allows the identification of the emitting atom and the composition of the surface analyzed (see Figure 1.38). In addition to these techniques, optically-based vibrational spectroscopy can allow the material composition to be determined through analysis of molecular vibrations (infrared and Raman spectroscopy).

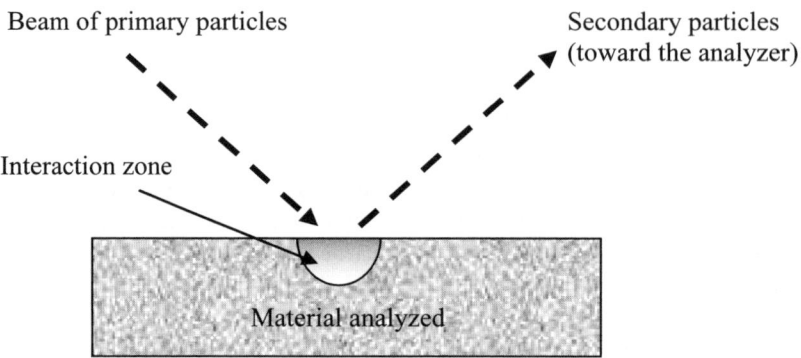

Figure 1.38. *Principle of spectroscopic surface analysis. Primary and secondary particles are electrons, photons or ions. The characterization of secondary particles (through their wavelength, speed, mass, etc.) enables us to define the nature of the emitting atom and determine the sample's composition*

There has been extensive research in surface analysis techniques [AGI 90, DAV 88, EBE 89]; here we provide a brief description of the most widely used techniques in the field of material science.

1.2.5.1. *Energy dispersive X-ray analysis*

– Primary particle: electron.

– Secondary particle: photon.

– Acronym: EDX.

– Characteristics of the technique:

- elemental analysis of materials to a depth of a few microns,

- possibility of generating X-ray images showing the distribution of elements,

- detection sensitivity in the range of 0.1–1 wt%, depending on the atomic number, and

- not well-suited for light element analysis ($Z < 11$).

POSSIBLE APPLICATIONS – Analysis of precipitation and segregation in metallurgical materials and the characterization of thick thermal oxide films and hard anti-wear and anti-corrosion coatings.

1.2.5.2. *X-ray photoelectron spectroscopy*

– Primary particle: photon.

– Secondary particle: electron.

– Acronym: ESCA, XPS.

– Characteristics of the technique:

- elemental analysis of materials,

- information about the environment of the atoms identified and the nature of chemical bonds, and

- analysis of a surface of a few mm^2 to a depth of 0.1–2 nm.

POSSIBLE APPLICATIONS – Analysis of oxide films and corrosion layers in order to determine the nature of chemical bonds of the main elements as well as that of impurities; analysis of passivation layers; characterization of the state of chemical bonds within polymers; and analysis of the core level and valence band.

1.2.5.3. *Auger electron spectroscopy*

– Primary particle: electron.

– Secondary particle: electron.

– Acronym: AES.

– Characteristics of the technique:

- elemental analysis of the top few-atom layers of solids. The depth of analysis is 2 to 3 monolayers (of the order of a nanometer),

- local analysis (primary electron beam diameter less than 10 nm); the use of an ion beam allows in-depth analysis alternating between analysis and ionic erosion (down to several microns). This is also made possible by photoelectron spectroscopy (XPS).

POSSIBLE APPLICATIONS – Analysis of adsorbed layers. Analysis of thermal or anodic oxide films. Characterization of thin deposits and analysis of segregated layers on surfaces or within grain boundaries.

1.2.5.4. *Glow discharge optical emission spectroscopy*

– Primary particle: ion.

– Secondary particle: photon.

– Acronym: GDOS.

– Characteristics of the technique:

- simultaneous elemental analysis of dozens of elements. Analysis of all elements, including hydrogen, possible with high sensitivity (10 ppm),

- sputtering analysis, a destructive technique which allows in-depth sample analysis,

- analysis over large areas (of the order of cm^2).

POSSIBLE APPLICATIONS – Study of diffusion gradients. Analysis of thin films or composition gradients. Identification of molecules adsorbed on a surface and characterization of coatings. Analysis of iron and steel materials.

1.2.5.5. *Rutherford backscattering spectroscopy*

– Primary particle: ion.

– Secondary particle: ion.

– Acronym: RBS.

– Characteristics of the technique:

- the analysis of the energy of backscattered helium ions after elastic collision with the sample's surface atoms allows the determination of their mass,

- several elements can be simultaneously analyzed over a depth of a few microns.

POSSIBLE APPLICATIONS – Analysis of thin films. Determination of concentration profiles. Suitable for all types of material.

1.2.5.6. *Secondary ion mass spectroscopy*

– Primary particle: ion.

– Secondary particle: ion.

– Acronym: SIMS.

– Characteristics of the technique:

- elemental and isotopic analysis of all elements in the periodic table,

- determination of concentration profiles of chemical elements to a depth of several microns using ionic sputtering,

- detection of trace elements (ppm),

- can be used to obtain ionic images showing elemental distribution.

POSSIBLE APPLICATIONS – Analysis of concentration profiles of bulk materials or thin layers. For example, in the semi-conductors industry (analysis of silicon, gallium arsenide or other semi-conductors; the aim is to control the dopant concentration profiles, but also to localize and quantify impurities). Also used to analyze metallurgical products heterogenity.

1.2.5.7. *Infrared spectrometry*

– Primary particle: photon.

– Secondary particle: photon.

– Acronym: IR spectrometry

– Characteristics of the technique:

- a method principally used for the analysis of organic materials,

- allows the determination of functional groups present in a material and the nature of the different bonds in which the carbon atoms are involved,

- an essentially qualitative technique,

- allows the analysis of gaseous, liquid and solid samples, even in very small quantities.

POSSIBLE APPLICATIONS – Analysis of plastics and lubricants such as oils or greases. Analysis of polymer-based coatings and thin films such as paints, varnishes or electrodeposited polymers.

Chapter 2

Tribology

2.1. Introduction

The word tribology is derived from the Greek word tribos, meaning friction. It was first used in the UK in 1966 to describe the scientific and technical domains studying friction, wear and lubrication.

Tribology is now a wide-ranging, multi-disciplinary field of study and research aimed at:

– reducing material wear and increasing the lifetime and reliability of mechanical and mechatronic systems; and

– controlling (or optimizing) friction (note that in some cases, e.g. vehicle brakes, we aim for maximum friction, whereas in others, e.g. skiing and surfing, the aim is to minimize its effects).

This dual objective requires an approach that combines contact mechanics with the physico-chemistry of surfaces and interfaces.

We will discuss these two aspects of tribology in this chapter, and present results that illustrate the coupling and interaction between mechanics and physico-chemistry in problems of friction and wear. However, before we enter into a detailed treatment of these effects, we first review a few basic notions of plasticity.

2.2. Elements of solid mechanics

2.2.1. *The stress vector*

When a surface dS with normal $\vec{n}$ (see Figure 2.1) is subject to a force $d\vec{F}$ at the point P, the stress vector $\vec{T}$ at P is defined by the relation:

$$d\vec{F} = \vec{T}dS \qquad [2.1]$$

The stress vector $\vec{T}$ is the force per unit area, written as $\vec{T}(P, \vec{n})$, corresponding to the stress vector at P acting on the face of normal $\vec{n}$. $\vec{T}$ can be decomposed into two parts:

$$\vec{T} = \vec{\sigma} + \vec{\tau} \qquad [2.2]$$

where $\vec{\sigma}$ is the normal stress and $\vec{\tau}$ the shear stress.

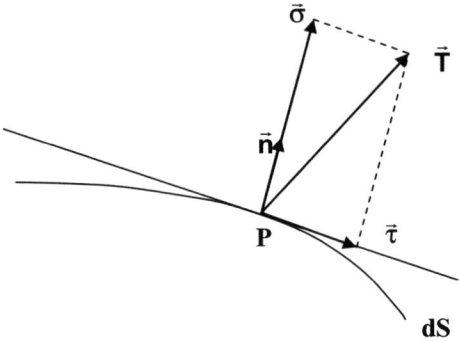

Figure 2.1. *The stress vector*

2.2.2. *The stress tensor*

To establish the state of stresses at a point P, it is sufficient to know the components of the stress tensor $\overline{\overline{\sigma}}$ at this point.

$$\overline{\overline{\sigma}} = \begin{pmatrix} \sigma_{xx} & \sigma_{xy} & \sigma_{xz} \\ \sigma_{yx} & \sigma_{yy} & \sigma_{yz} \\ \sigma_{zx} & \sigma_{zy} & \sigma_{zz} \end{pmatrix} \qquad [2.3]$$

If we consider a parallelepiped of sides dx, dy, dz in equilibrium as shown in Figure 2.2, each component σ_{ij} represents the stress parallel to the direction i applied to the face with normal j. σ_{xx}, σ_{yy} and σ_{zz} are tension stresses, whereas the six other components of the tensor are shear stresses. They are often written as τ instead of σ, noted as τ_{xy}, τ_{yx}, τ_{xz}, τ_{zx}, τ_{yz}, τ_{zy}.

Because the body is isotropic and in equilibrium, we can write:

$$\sigma_{ij} = \sigma_{ji}$$

The tensor $\overline{\overline{\sigma}}$ therefore reduces to six components, and can be diagonalized in the following form:

$$\overline{\overline{\sigma}} = \begin{pmatrix} \sigma_I & 0 & 0 \\ 0 & \sigma_{II} & 0 \\ 0 & 0 & \sigma_{III} \end{pmatrix} \qquad [2.4]$$

where σ_I, σ_{II} and σ_{III} are the principal stresses expressed in this new orthonormal frame and the new axes are referred to the principal directions.

In the principal frame, the stress tensor therefore reduces to three components σ_I, σ_{II} and σ_{III}.

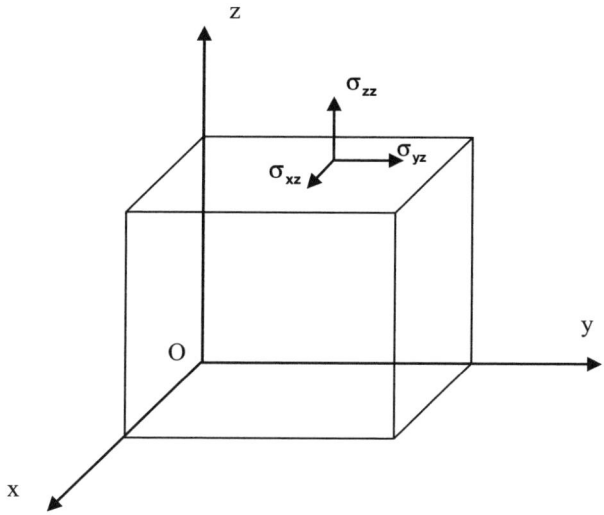

Figure 2.2. *Components of the stress tensor*

A given state of stress described in the principal frame by the tensor $\bar{\bar{\sigma}}$ can therefore be decomposed into two terms

$$\begin{pmatrix} \sigma_I & 0 & 0 \\ 0 & \sigma_{II} & 0 \\ 0 & 0 & \sigma_{III} \end{pmatrix} = \begin{pmatrix} \sigma_m & 0 & 0 \\ 0 & \sigma_m & 0 \\ 0 & 0 & \sigma_m \end{pmatrix} + \begin{pmatrix} \sigma_I - \sigma_m & 0 & 0 \\ 0 & \sigma_{II} - \sigma_m & 0 \\ 0 & 0 & \sigma_{III} - \sigma_m \end{pmatrix} \quad [2.5]$$

The first term is referred to as the spherical part of the tensor $\bar{\bar{\sigma}}$ and it corresponds to a hydrostatic pressure. The second term represents the deviatoric stress and corresponds to the shear component of tensor $\bar{\bar{\sigma}}$.

2.2.3. Yield criteria

Yield criteria are expressed as a series of equations defining particular conditions on the components σ_{ij} of the stress tensor. They are noted in the form:

$$f(\sigma_{ij}) = 0 \quad [2.6]$$

where f is a known function. When this condition is satisfied, yield will occur.

Equation [2.6] can be expressed as a function of the principal stresses in terms of a new function f':

$$f'(\sigma_I, \sigma_{II}, \sigma_{III}) = 0 \quad [2.7]$$

2.2.3.1. The Tresca criterion

The Tresca criterion is a maximum shear stress criterion based on the notion that yielding occurs when the maximum shear stress attains a threshold value.

Taking $\sigma_I \geq \sigma_{II} \geq \sigma_{III}$, we can write this criterion:

$$\tau_{max} = \frac{\sigma_I - \sigma_{III}}{2} \geq k = \frac{Y}{2} \quad [2.8]$$

When τ_{max} is greater than or equal to the yield shear stress for a state of pure stress k, yielding occurs. Y gives the yield stress obtained from a tensile (or compression) test.

2.2.3.2. *The von Mises criterion*

This criterion is based on the notion that plastic deformation occurs when the stored distortion strain energy attains a critical value.

Using Y and k defined above, the von Mises criterion is expressed in the principal frame by:

$$(\sigma_I - \sigma_{II})^2 + (\sigma_{II} - \sigma_{III})^2 + (\sigma_{III} - \sigma_I)^2 = 2Y^2 = 6k^2 \qquad [2.9]$$

The two criteria have been superimposed in Figure 2.3, and we can note that the maximum separation (E) between the two is equal to:

$$E = \sqrt{\frac{2}{3}}Y - \frac{Y}{\sqrt{2}}, \qquad [2.10]$$

a maximum difference of around 15%.

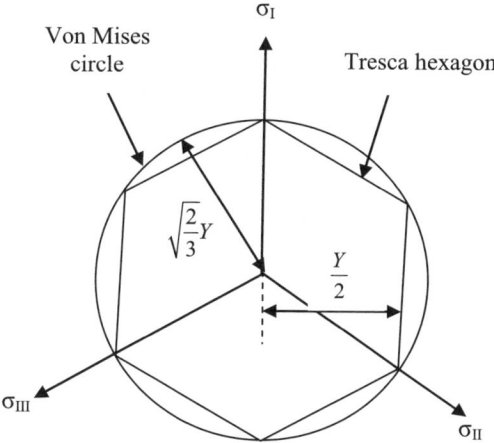

Figure 2.3. *Representation of the Tresca and von Mises criteria in the principal frame*

2.3. Elements of contact mechanics [JOH 85, WIL 94]

2.3.1. *Hertz contact theory*

Figure 2.4 shows three different contact geometry configurations. In the case of elastic contact, Hertz's theory allows us to describe the stress distribution on and below the surface.

54   Materials and Surface Engineering in Tribology

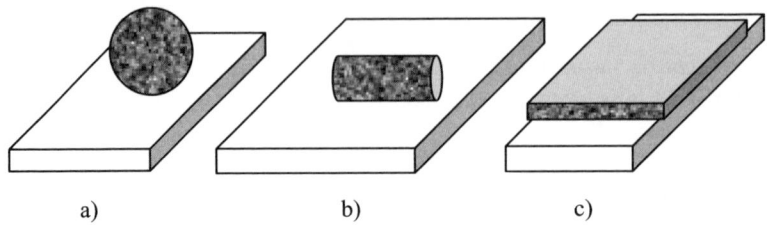

Figure 2.4. *Different types of contacts:*
a) point; b) linear; c) surface

We take a simple case of point contact, illustrated by a sphere of radius R pushing on a plane surface under the action of a force F (see Figure 2.5).

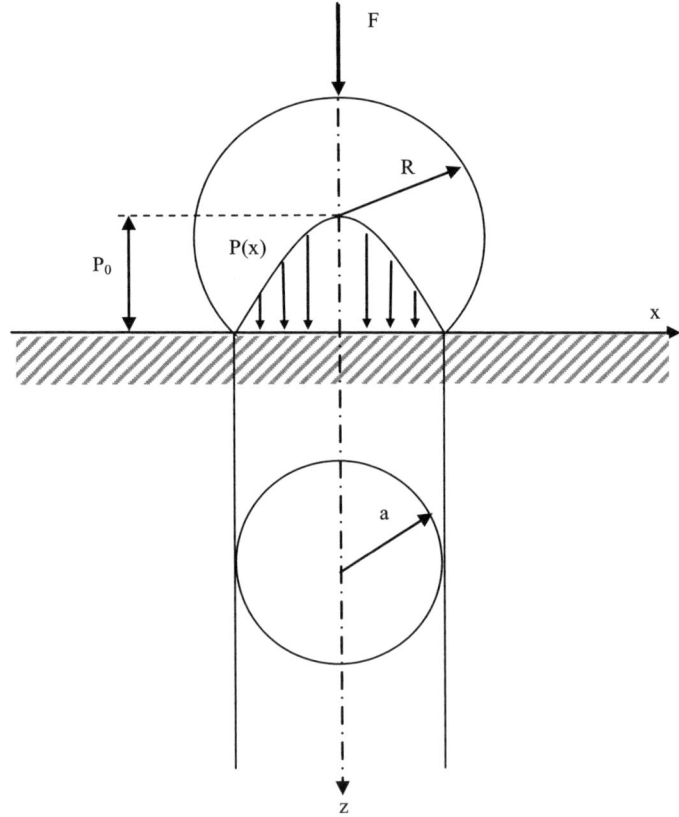

Figure 2.5. *Sphere-plane contact*

The load is distributed on a circular contact area of radius a:

$$a = \left(\frac{3FR}{4E}\right)^{\frac{1}{3}} \quad [2.11]$$

where E is defined by:

$$\frac{1}{E} = \frac{1-v_1^2}{E_1} + \frac{1-v_2^2}{E_2} \quad [2.12]$$

where E_1 and E_2 are Young's moduli and v_1 and v_2 correspond to Poisson's coefficients for the sphere and plane, respectively.

The pressure distribution on the contact area can be written:

$$P = P_0\sqrt{1-\left(\frac{r}{a}\right)^2} \quad [2.13]$$

where r can take x-values ranging from a to –a (see Figure 2.5).

This expression shows that the contact pressure is zero at the edges and maximum at the center. The maximum pressure P_0 (also called the Hertz pressure) is given by:

$$P_0 = \frac{3F}{2\pi a^2} \quad [2.14]$$

and it is 1.5 times greater than mean contact pressure P_m:

$$P_m = \frac{F}{\pi a^2} \quad [2.15]$$

Depending on the value of P_m relative to the yield stress Y, the deformation will be elastic, elastoplastic or plastic:

– for $0 \leq P_m < 1.1Y$, we have truly elastic deformation;

– $P_m = 1.1Y$ corresponds to the threshold for initial plastification; and

– $P_m = 3Y$ corresponds to complete plastification.

Figure 2.6 shows the surface distribution of stress σ_{xx} (in the contact plane). The stress is compressive within the contact region and tractive at its edges.

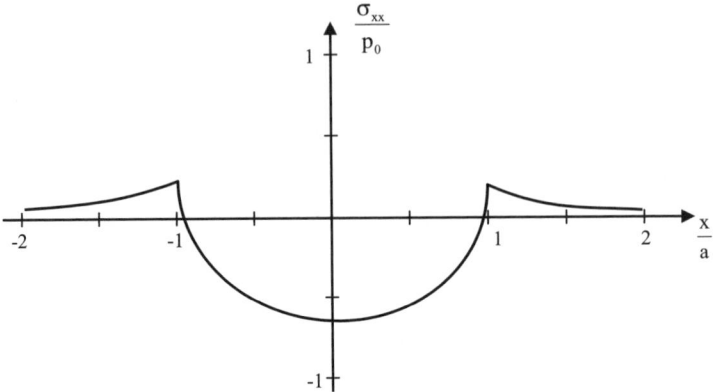

Figure 2.6. *Surface stress in a sphere-plane contact*

Under the surface, along the z axis, the shear stress is maximal; $\tau_{max} = 0.31\ P_0$ at the point $z = 0.48a$, as shown in Figure 2.7. It is at this point that the initial yielding occurs.

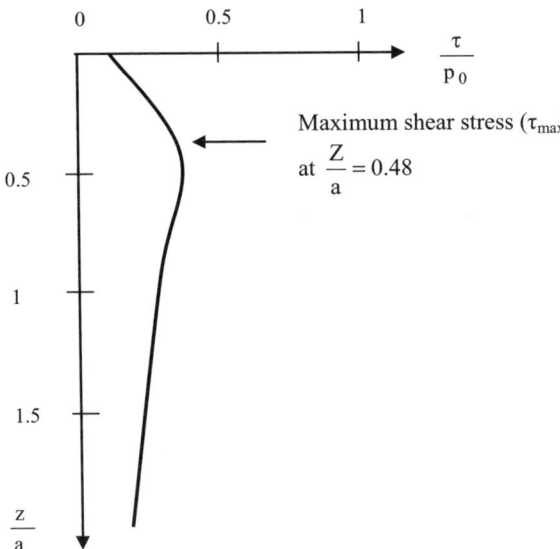

Figure 2.7. *Below-surface shear stress in the case of a sphere-plane contact*

When a sphere slides on a plane, the distribution of stresses is significantly modified. Figure 2.8 shows the new profile of the stress σ_{xx} which loses its symmetry and becomes a compressive stress at the front edge of the contact and a tensile stress at the trailing edge of the contact. It is also clear that its amplitude grows sharply as a function of the coefficient of friction μ and that it can lead, in the case of brittle materials, to the development of characteristic, crescent-shaped cracks (see Figure 2.9).

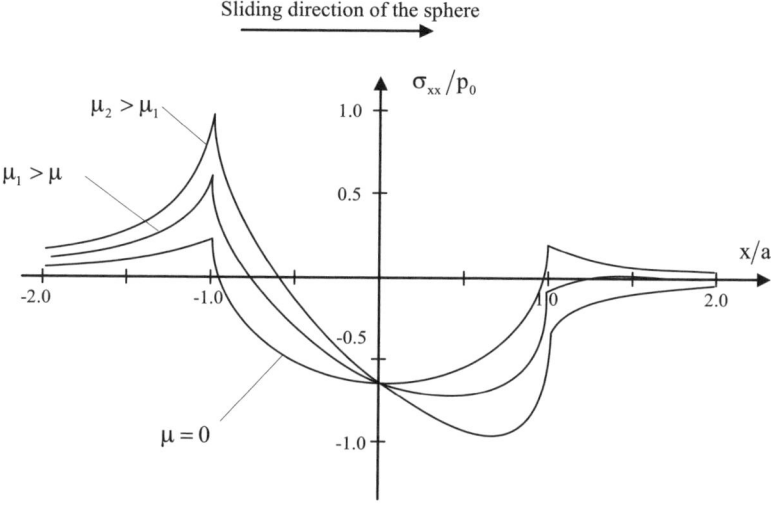

Figure 2.8. *Sliding of a sphere on a plane: alteration of surface stresses relative to friction coefficient μ (static case i.e. μ = 0 and evolution after displacement with $\mu_2 > \mu_1 > \mu$)*

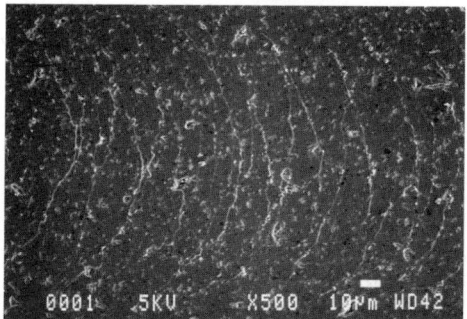

Figure 2.9. *Crescent-shaped cracks on a SiC surface after friction with a SiC sphere moved from right to left*

2.3.2. *The contact area*

In the case of two plane rough surfaces placed in contact and subject to a normal load F, contact occurs at points at the tips of the asperities, as shown in Figure 2.10a. If the normal load is increased and replaced by a new force F' > F, new points of contact appear and the true contact area increases (see Figure 2.10b). This can be written $A_r = \Sigma A_{r(i)}$ (where $A_{r(i)}$ is the contact area between two asperities) and is very small with respect to the apparent surface contact area ($A_a = XY$), as shown in Figure 2.11. The A_a/A_r ratio depends on the distribution of asperities, on the normal load and on the yield stresses of the materials.

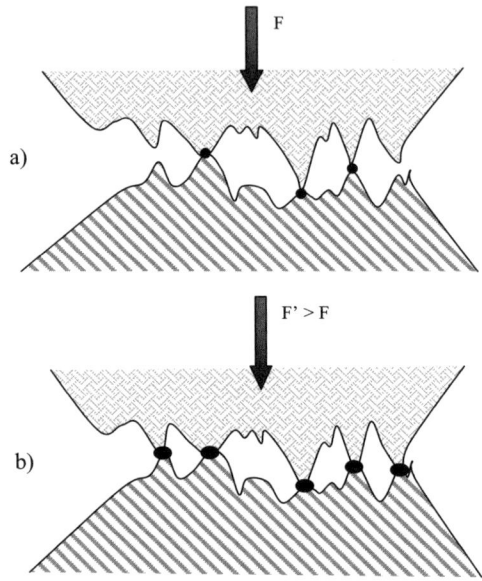

Figures 2.10. *Two rough surfaces in contact under a) a load F and under b) F' > F. When the load increases, the number of asperities in contact also increases*

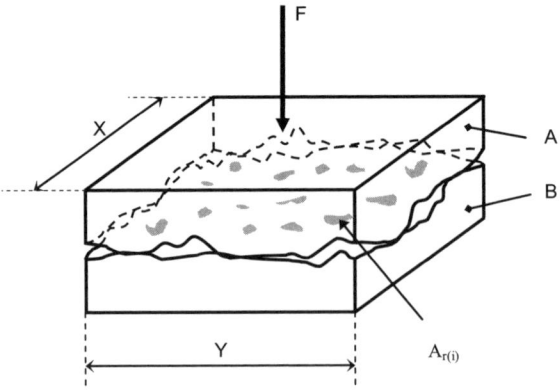

Figure 2.11. *Contact areas between two solids A and B: the apparent contact area ($A_a = XY$) and the true contact area ($A_r = \Sigma A_{r(i)}$)*

The apparent contact area is usually between 100 and 10,000 times greater than the true contact area [BRI 92]. Note that during friction, this contact area changes, both in terms of its total extent as well as its chemical and structural composition.

2.3.3. *Plastification of asperities*

For the case of a Gaussian distribution of asperities, the contact pressure is expressed using [GRE 66]:

$$P = 0.3 \left[\frac{E}{(1-v^2)} \right] \sqrt{\frac{\sigma}{\beta}} \qquad [2.16]$$

where β is the mean radius of curvature of the (assumed spherical) asperities and σ the standard deviation of their height distribution.

A plasticity index ψ has been introduced in order to determine the limit between elastic and plastic deformation of the asperities in contact. In the case of a Gaussian distribution of asperities, this index is expressed as the product of two factors: the first defines the mechanical properties of the materials whereas the second accounts for the surfaces' topographies

The plasticity index ψ is written:

$$\psi = \left[\frac{E}{H(1-v^2)} \right] \sqrt{\frac{\sigma}{\beta}} \qquad [2.17]$$

where E, H and ν represent the Young's modulus, hardness and Poisson's coefficient for the softest material, respectively.

The nature of the deformation of the asperities depends on the value of Ψ as follows:

– ψ ≥ 1: plastic contact;
– 1 > ψ ≥ 0.6: elasto-plastic contact; and
– ψ < 0.6: elastic contact.

2.3.4. *Adhesive contact*

The contact area, as calculated above based on Hertz's theory (equation [2.11]), does not take into account the forces of molecular attraction (mainly Van der Waals forces) which can noticeably modify its dimensions. Two models can be used to calculate the new contact area (see Figure 2.12).

The first is the *Johnson, Kendall and Roberts* or *JKR* model [JOH 71]. It is based on an energetic criterion and supposes the forces of attraction to be confined within the contact area i.e. are zero outside it (Figure 2.12b).

The second is the *Derjaguin, Muller and Toporov* or *DMT* model [DER 75]. It takes into account the attractive forces outside the contact area but supposes an unmodified sphere-plane contact profile (Hertz-type contact) (Figure 2.12c).

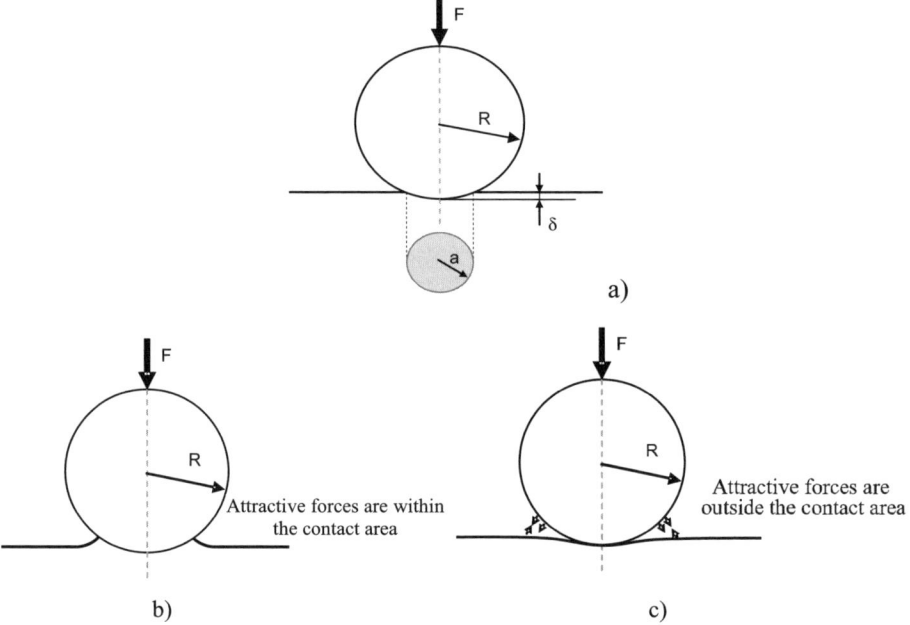

Figure 2.12. *a) Sphere-plane contact showing the penetration δ of the sphere into the plane; b) sphere-plane contact showing attractive forces confined to the contact area (JKR model); c) Hertz contact with attractive forces outside of the contact area (DMT model)*

Table 2.1 summarizes the main characteristics of the three types of sphere-plane contacts: Hertz, JKR and DMT. Here, F_{ad} refers to the adhesive force required to break the contact, a is the radius of the contact area under load F (a_0 under zero load), δ is the total deformation (see Figure 2.12a), R is the sphere radius, F is the normal applied load, W is the adhesive force between the sphere and the plane and K is given by the relation:

Tribology 61

$$\frac{1}{K} = \frac{3}{4}\left(\frac{1-v_1^2}{E_1} + \frac{1-v_2^2}{E_2}\right) \quad [2.18]$$

	Hertz Contact	JKR Model	DMT Model
F_{ad}	0	$\frac{3}{2}\pi WR$	$2\pi WR$
A	$\left(\frac{R}{K}F\right)^{\frac{1}{3}}$	$\left(\frac{R}{K}\left(F + 3\pi RW + \sqrt{6\pi RWF + (3\pi RW)^2}\right)\right)^{\frac{1}{3}}$	$\left(\frac{R}{K}[F + 2\pi RW]\right)^{\frac{1}{3}}$
a_0	0	$\left(\frac{6\pi WR^2}{K}\right)^{\frac{1}{3}}$	$\left(\frac{2\pi WR^2}{K}\right)^{\frac{1}{3}}$
δ	$\frac{a^2}{R}$	$\frac{a^2}{R} - \frac{2}{3}\sqrt{\frac{6\pi Wa}{K}}$	$\frac{a^2}{R}$

Table 2.1. *Some characteristic values for the sphere-plane contact*

The JKR model is mainly valid for the case of high surface energy materials with large contact areas (soft materials or large radius spheres, typically of a few mm), whereas the DMT model is mostly valid in the case of hard materials of small radius spheres and low surface energy.

Figure 2.13 shows the evolution of the contact area between a sharp tip and a plane surface as a function of the normal applied load. When this load is zero, the contact area also becomes zero in the absence of surface forces (Hertz contact). When surface forces are acting (JKR or DMT) we note that, even when the applied load is zero or negative, the contact area always has a non-zero value.

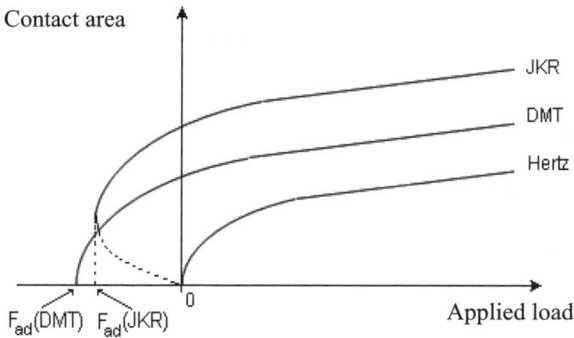

Figure 2.13. *Variation of the contact area as a function of the applied load: a comparison of the three models Hertz, DMT and JKR*

Several authors have reported that the adhesive force due solely to surface forces can be sufficient to induce plastification of the indented surface. Values ranging from a few micronewtons to several tens of micronewtons have been recorded in a vacuum between metallic surfaces [CHO 80, PAS 84, POL 77].

If γ corresponds to the surface energy and we have two identical materials 1 and 2 (see section 1.2.3), we can write:

$$\gamma_1 = \gamma_2 = \gamma \text{ and } \gamma_{12} = 0 \qquad [2.19]$$

which allows us to write:

$$W = \gamma_1 + \gamma_2 - \gamma_{12} = 2\gamma \qquad [2.20]$$

Using equation [2.20] and the expressions for F_{ad} (see Table 2.1), we therefore have:
– for the JKR model:
$$F_{ad} = 3\pi\gamma R \qquad [2.21]$$
– and for the DMT model:
$$F_{ad} = 4\pi\gamma R \qquad [2.22]$$

It is therefore possible to determine the surface energy of the material by carrying out experimental measurements of F_{ad} (see section 2.5.2). However, this is only possible if the separation force, which is negative here, is applied sufficiently slowly to the contact in order to ensure reversible behavior. When the speed of separation increases, visco-elastic losses can occur in the material (as has been observed with rubber for example), which result in γ values as high as 100 to 1,000 times those predicted by the laws of thermodynamics [PAS 84].

Finally, we note that the adhesive contact also depends on temperature and duration of the contact between the materials [MAU 78, MAU 84].

2.4. Friction

2.4.1. *The coefficient of friction*

Consider a solid parallelepiped on a horizontal plane subject to a normal load F_n. We now progressively apply a force F parallel to the plane in order to set it into motion and increase its speed (v) from 0 to V (see Figure 2.14). This displacement will result in a friction force F_t in the plane, in the opposite direction to the sliding motion and opposing it. Figure 2.15a shows the temporal evolution of this force.

Knowing F_n and having determined F_t experimentally, we can then calculate the coefficient of friction μ which is defined by:

$$\mu = \frac{F_t}{F_n} \qquad [2.23]$$

We note that the coefficient of static friction (μ_s with $F_t = F_{t(s)}$) is distinct from the coefficient of dynamic friction (μ_d, with $F_t = F_{t(d)}$) as illustrated in Figure 2.15a:

– $F_{t(s)}$ is the maximum force to be applied to set the solid into motion; and

– $F_{t(d)}$ is the force applied to maintain this motion.

We often observe the evolution of the frictional force shown in Figure 2.15b. This corresponds to the *stick-slip* phenomenon which results from a series of adhesions and breakings of the contact at the contact points between opposing surfaces (see Figure 2.11).

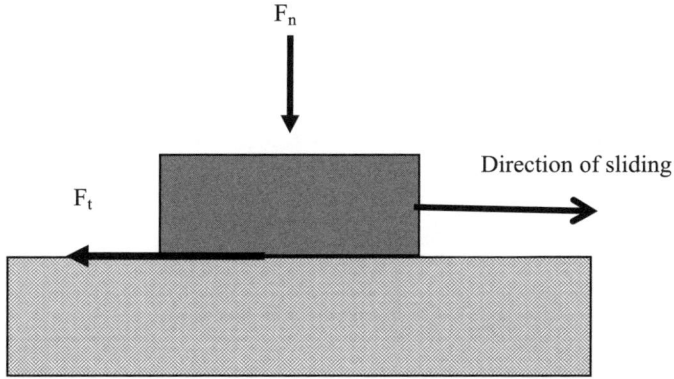

Figure 2.14. *Definition of the tangential force F_t.*
Interfacial friction is the origin of a solid's resistance to sliding

64   Materials and Surface Engineering in Tribology

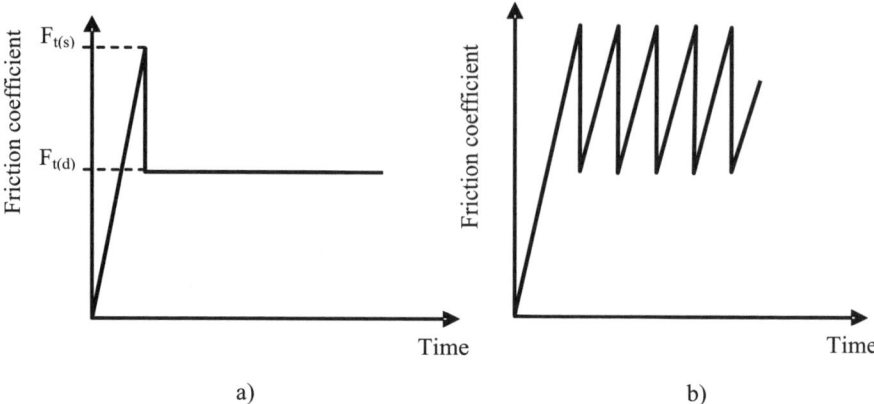

Figure 2.15. *Evolution of the friction force through time:
a) sliding without* stick-slip*; b)* stick-slip *sliding*

Depending on the nature of the materials and the experimental conditions, several different types of behavior can be observed as illustrated in Figure 2.16. They result from chemical, topographical or structural modifications of the surfaces in sliding contact and can include oxidation, allotropic phase transformation, amorphization, crystallization, diffusion, melting, polishing or removal of material.

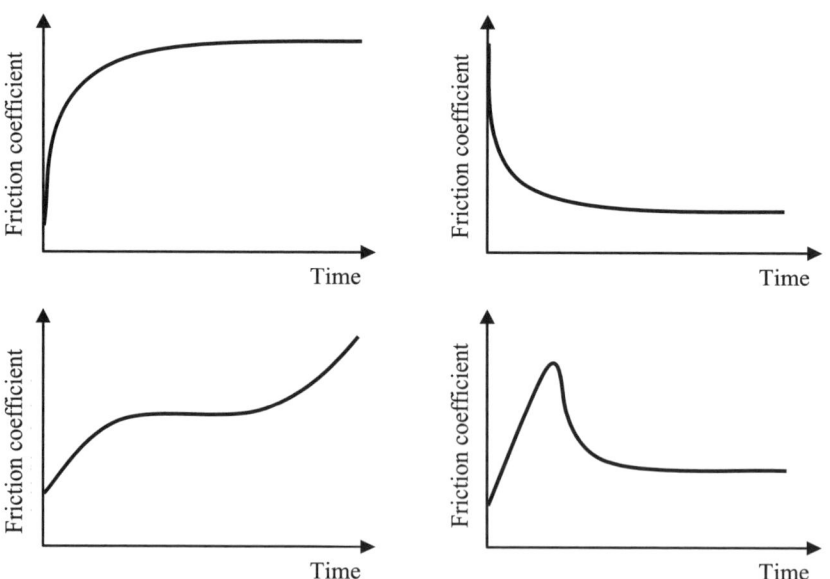

Figure 2.16. *Examples of evolution of the friction force as a function of time*

In Figure 2.17, we illustrate the case of two metallic surfaces in sliding contact and the subsequent formation and shearing of a junction of area A, showing the transfer of matter from the softest (B) to the hardest material (A).

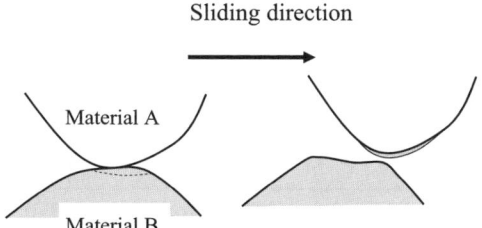

Figure 2.17. *Adhesion with transfer of matter*

Under these conditions, if we rewrite equation [2.23], replacing F_t with the product $A\tau$ and F_n with the product AP_m, we obtain:

$$\mu = \frac{A\tau}{AP_m} = \frac{\tau}{P_m} \qquad [2.24]$$

In this expression, τ is the shear stress and P_m is the mean contact pressure.

If we adopt the von Mises flow criterion since plastification has occurred, we can write:

$$P_m = 3Y \qquad [2.25]$$

and:

$$\tau = \tau_c = \frac{Y}{\sqrt{3}} \qquad [2.26]$$

where Y is the yield stress and τ_c the critical shear stress of the softest material.

If we then rewrite equation [2.24], taking into account equations [2.25] and [2.26], we obtain: $\mu \approx 0.2$.

Although this is a simple model it can yield acceptable values for μ, even if they are slightly low compared to mean values obtained for metal–metal pairs. Table 2.2 presents some values for coefficients of friction of different materials.

Materials	Friction coefficients
Metal–metal in vacuum (< 10^{-7} Pa)	> 3
Meta l–metal in air	0.2 to 1.5
Polymer–polymer in air	0.05 to 1
Metal–polymer in air	0.05 to 0.5
Metal–ceramic or ceramic–ceramic in air	0.2 to 0.5
Metal–metal with solid lubricant (PTFE, MoS2, graphite)	0.05 to 1
Metal–metal with oil lubricant (outside hydrodynamic regime)	0.1 to 0.2
Lubrication in the hydrodynamic regime	0.001 to 0.005

Table 2.2. *Some values for friction coefficients*

We recall that friction is extremely sensitive to the surface cleanliness and to the ambient environment in which the materials are placed. In ultra-vacuum, there are no adsorbed species on the surfaces and they are very reactive. This results in strong adhesive contact and therefore high coefficients of friction. Conversely, in air or in a reactive atmosphere, the surface is covered with a layer composed of reaction products with the atmosphere as well as additional adsorbed species or contaminants. This layer acts as a lubricant and considerably reduces the coefficient of friction (see section 1.2.1).

Experiments carried out in an ultra-vacuum chamber (less than 10^{-7} Pa) have allowed measurement of friction coefficients exceeding 10 for many metal–metal pairs (e.g. 12 for Fe/Fe, 40 for Zr/Zr and 60 for Ti/Ti). The introduction of increasing amounts of oxygen (or other reactive gases) in the chamber results in a systematic reduction of the friction coefficient. The effect is enhanced as the metal presents stronger affinity to the gas introduced [BUC 81].

2.4.2. *Tribometers*

Tribological tests are performed with tribometers which can operate in air or in a controlled atmosphere, with or without lubrication. Figure 2.18 shows an example of a ball-on- disk laboratory tribometer with a reciprocating motion. The parameters imposed are generally the applied load, the sliding speed and the environmental conditions (humidity and controlled atmosphere i.e. nature and pressure of gases introduced). The quantities measured are generally the friction force, the surface temperature, the contact resistance and the wear.

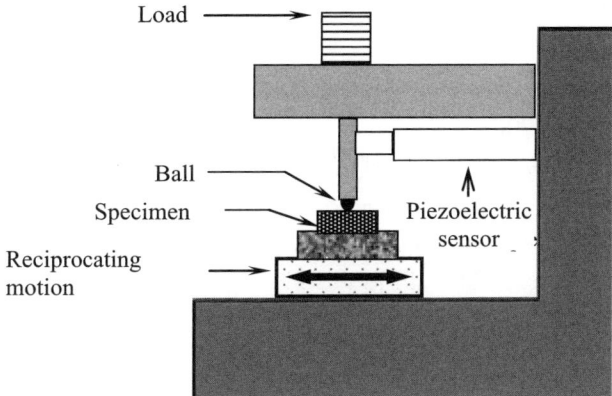

Figure 2.18. *Ball-on- disk tribometer*

Depending on the particular application, tests can be conducted using a number of different contact geometries (see Figure 2.19).

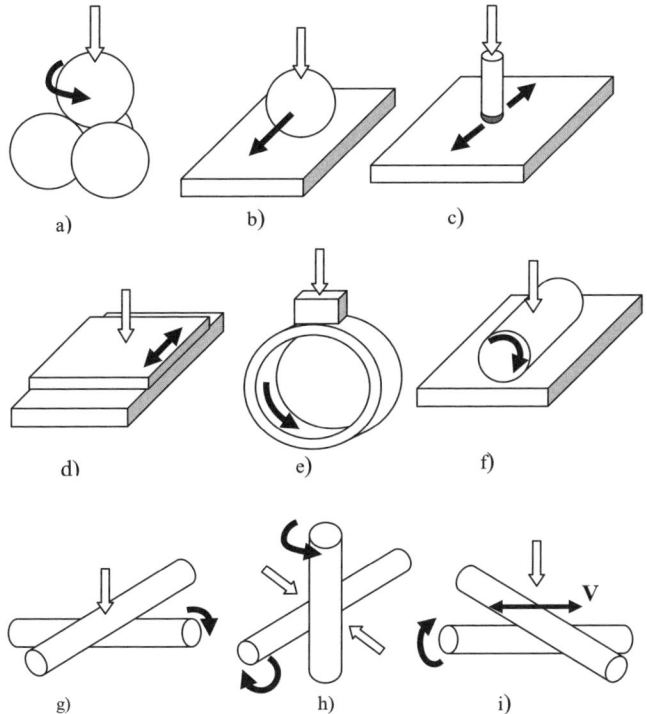

Figure 2.19. *Variety of contact geometry in tribometers: a) four-ball; b) sphere/plane; c) pin-on- disk; d) plane/plane; e) plane/cylinder; f) cylinder/plane; g–i) cylinder/cylinder*

2.4.3. Laws and theories of friction

Historically, it was Leonardo da Vinci who first attempted to provide a scientific explanation for friction (circa 1500) and who introduced the notion of friction coefficient. In the 1700s, Guillaume Amontons formulated the first two laws of friction which still bear his name today:

– the friction force is proportional to the normal applied load; and
– the friction force is independent of the apparent contact area.

These laws have been verified with mostly metals. However, with very hard materials or highly elastic materials (such as rubber), experimental results have not agreed with theoretical predictions [BOW 50]. It also must be noted that even for metals, when the contact pressure is lower than the plastic flow threshold, we observe behavior which is the opposite to that described by the first law. It was Charles-Augustin Coulomb who introduced a distinction between static and dynamic friction coefficients, and formulated the third law of friction in the 1780s:

– the friction coefficient is independent of the sliding speed.

All three of these rules have been verified in many cases, yet they should be used with a degree of prudence as they do not apply to all materials regardless of the environment and types of stress. This is particularly true when sliding speeds are too high or when too large a range of loads is used [BLAU 95].

It was in the 1950s that a microscopic approach was introduced, based on the formation and rupture of junctions at the contact points between opposing surfaces [BOWD 50, BOWD 64]. Under the combined effects of the applied load and the sliding speed, the interfacial temperature increase can lead to the growth of numerous junction points between the solids. When these junctions are weak, shear occurs within but with little or no transfer of matter. Conversely, when they are strong, shear occurs in the softest material which is transferred onto the harder material. For example, this is what happens when a soft copper or copper-based sphere is rubbed against a hard steel surface (see Figure 2.20).

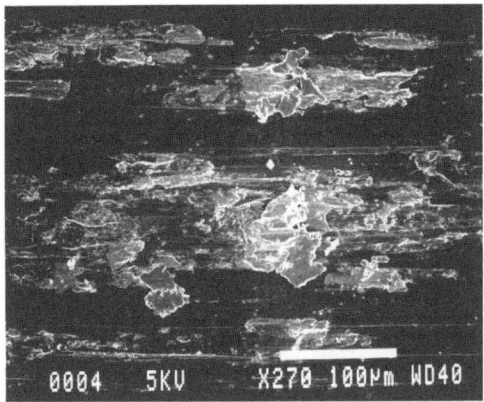

Figure 2.20. *Bronze particles (light) transferred onto a steel surface (dark) by friction*

In order to explain friction, a theory based on the overlapping of surface asperities was also proposed but was soon abandoned when it was established that carefully-polished surfaces could still have high friction coefficients.

Today, however, we know that depending on the nature of materials and the conditions of the surfaces in contact, there exists an optimal value of roughness that minimizes the friction. The smoother the surface and the lower the roughness, the greater the true contact surface will be and, consequently, the greater the adhesive friction component will be (see equation [2.27]). Conversely, the rougher the surface and the sharper and the more numerous the asperities, the greater the plastic deformation component will be. Between these two extremes, there exists an optimal roughness associated with a low friction coefficient (see Figure 2.21).

It should be noted that, for a given surface state, the evolution of the friction coefficient relative to the normal applied load is similar to that shown in Figure 2.21. There exists an optimal load that minimizes the friction coefficient [MYS 97].

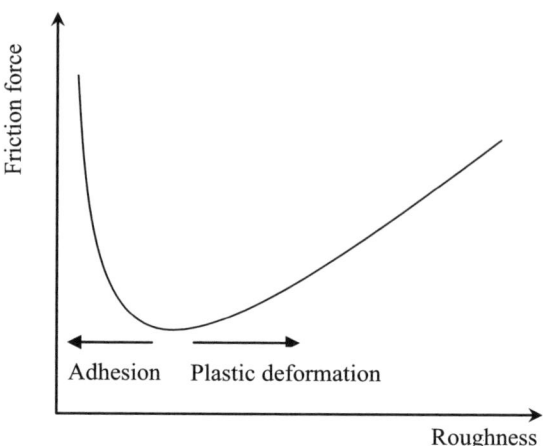

Figure 2.21. *Impact of roughness on friction: definition of optimal roughness. The evolution of friction relative to the applied load can be illustrated by a curve of similar shape*

It is widely known that the friction force consists of two terms and can be expressed as:

$$F_t = F_d + F_a \qquad [2.27]$$

The term F_d is due to the deformation of asperities and to the ploughing of the softest surface by the hardest surface (plastic deformation component), whereas F_a is the force needed to shear the adhesive junctions between opposing materials (adhesive component). For the case of polymers, friction also induces visco-elastic losses which constitute, in most cases, the predominant factor [BRISCOEE 98].

During the friction process, only part of the dissipated energy is due to the wear of the opposing materials. Indeed, [BRISCOEB 92] gives the following as a general guide to the forms of energy into which "mechanical energy is converted during friction:

– heat energy, causing an increase in the temperature of rubbing bodies;
– acoustic energy, producing audible effects;
– optical energy, including the full spectral range;
– electric energy, responsible for generating electrostatic charge;
– mechanical energy, causing wear of contacting bodies; and
– mechanical energy (or entropy), causing further comminution of wear particles".

Tribology 71

The relative contribution of these different forms of mechanical energy expenditure will differ according to the nature of the opposing materials, the surrounding environment and the contact conditions. There is therefore no correlation between wear and friction, and it is indeed possible to observe significant friction without any surface wear. Wear can therefore only be correlated with a fraction of the energy used in the wear process itself.

2.5. Nanotribology

2.5.1. *Surface forces*

Surface forces play an important part in friction and their impact is particularly noticeable with microcontacts, especially when the forces involved are low (from a few micronewtons to a few hundreds of micronewtons).

These forces are classified into three types:
- electrostatic forces;
- capillary forces; and
- Van der Waals forces.

2.5.1.1. *Electrostatic forces*

These forces result from Coulomb interactions between pre-charged surfaces or from surfaces which become charged through friction (triboelectricity). They can be described using the classic equations of electrostatics.

2.5.1.2. *Capillary forces*

The humidity content of ambient air normally ranges from 30 to 60%. Under these conditions, a film of water forms on the surface of materials and induces a new adhesive force known as the capillary force.

Figure 2.22 illustrates the formation of a meniscus between two solids – a sphere of radius R and a plane – separated by a liquid film. If the interaction between the solids is negligible (as is the case when D is large relative to the range of action of the surface forces), and if the sphere's radius R is also large relative to the sphere/plane distance, then the capillary force is given as a function of the interfacial tension γ_{LV} between liquid/vapor by [DEGE 05, GEO 00]:

$$F = 4\pi R \gamma_{LV} \cos\theta \qquad [2.28]$$

This shows that the force reaches a maximum (F_{max} for $\theta = 0$) corresponding to a capillary force that acts in the vertical direction (see Figure 2.23) [DEGE 05]. This

72   Materials and Surface Engineering in Tribology

result is related to equation [1.17] which shows that the liquid/solid adhesive energy is greatest for θ = 0.

Finally, we note that if the solids are placed in a dry environment or immersed in a liquid, then the capillary force is zero.

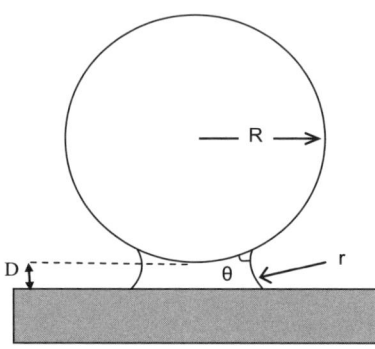

Figure 2.22. *Formation of a meniscus between two surfaces separated by a liquid film*

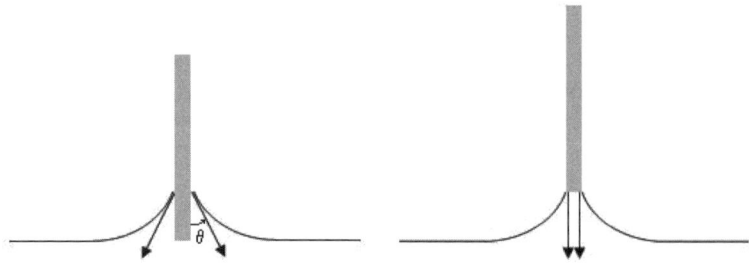

Figure 2.23. *Variations in the capillary force when an object dipped in a liquid is withdrawn. The capillary force reaches a maximum when θ = 0. Figure adapted from [DEGE 05]*

2.5.1.3. *Van der Waals forces*

Van der Waals forces result from dipole–dipole interactions, either between polar molecules or between polar and neutral molecules (induced dipoles). They can also arise from dispersive effects (London dispersion forces) due to electron motion which causes the appearance of instantaneous dipoles with very short lifetimes. In the case of solid–solid contacts, these forces can be observed by monitoring the evolution of the force of interaction between a sharp tip and a plane surface slowly brought into contact in a vacuum (see Figure 2.24).

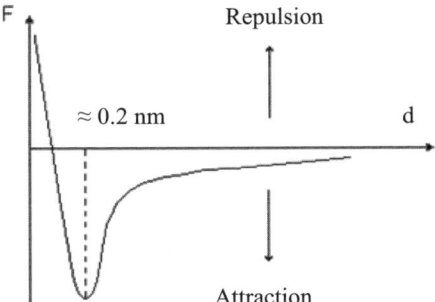

Figure 2.24. *Evolution of the force of interaction relative to the distance between a sharp tip and a surface*

When the two surfaces are sufficiently close they are attracted through the effect of Van der Waals forces. However, when the distance becomes less than about 0.2 nm (equilibrium distance D_0), nuclear repulsion becomes dominant. We note that the intermolecular attraction can be stronger due to the generation of genuine chemical bonds such as covalent or metallic bonds. On the other hand, when the surfaces are immersed in a liquid, new forces develop which determine the nature and the strength of the interactions [BHU 05]. Van der Waals forces can be expressed as a function of the contact geometry, using equations [2.29–2.31] (Figure 2.25).

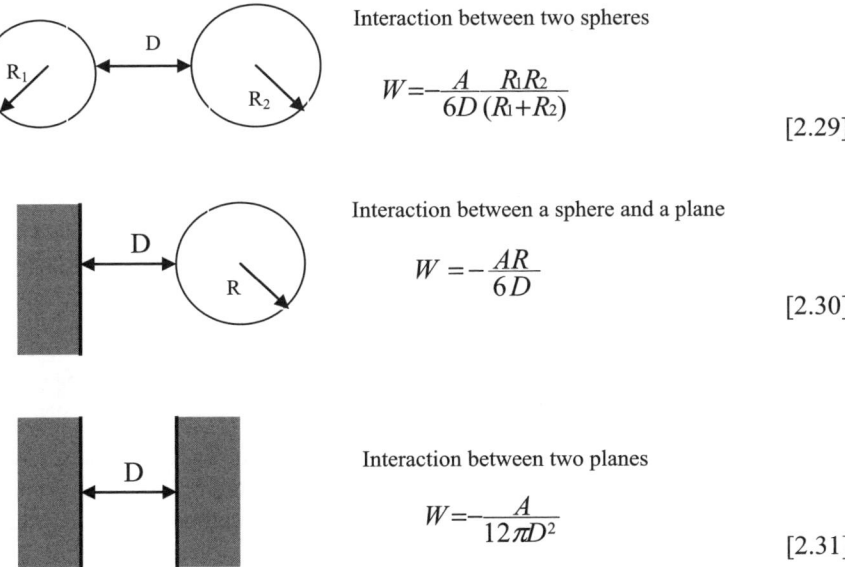

Interaction between two spheres

$$W = -\frac{A}{6D}\frac{R_1 R_2}{(R_1 + R_2)} \quad [2.29]$$

Interaction between a sphere and a plane

$$W = -\frac{AR}{6D} \quad [2.30]$$

Interaction between two planes

$$W = -\frac{A}{12\pi D^2} \quad [2.31]$$

Figure 2.25. *Interaction energy (W) between two solids of different geometry*

In these expressions, A represents Hamaker's constant, a complex function of the dielectric (dielectric permittivity ε) and optical (refractive index n) characteristics of the materials under study and the medium separating them.

If we consider two media 1 and 2 interacting across a medium 3, Hamaker's constant can then be expressed:

$$A = \frac{3}{4}kT\left(\frac{\varepsilon_1-\varepsilon_3}{\varepsilon_1+\varepsilon_3}\right)\left(\frac{\varepsilon_2-\varepsilon_3}{\varepsilon_2+\varepsilon_3}\right) + \frac{3h\nu_e}{8\sqrt{2}} \frac{(n_1^2-n_3^2)(n_2^2-n_3^2)}{\sqrt{(n_1^2+n_3^2)(n_2^2+n_3^2)}\left(\sqrt{(n_1^2+n_3^2)}+\sqrt{(n_2^2+n_3^2)}\right)} \quad [2.32]$$

where ε_i is the dielectric permittivity; n_i the refractive index of medium i; k and h are Boltzmann's and Planck's constants, respectively; T is the absolute temperature and ν_e is a frequency of the order 3×10^{15} s^{-1} [CAP 99, ISR 99]. Hamaker's constant usually takes values ranging between 10^{-19} and 10^{-20} J.

The expressions given in Figure 2.25 clearly show that, in all cases, the interaction energy is proportional to Hamaker's constant.

Consider two identical planes positioned at an equilibrium distance D_0 (state 1) and separated by a sufficient distance that they do not interact (state 2). If only Van der Waals forces are acting, the variation of the system energy will be:

$$\Delta W = W_{state\ 2} - W_{state\ 1} = 0 - \left(-\frac{A}{12\pi D_0^2}\right) = \frac{A}{12\pi D_0^2} \quad [2.33]$$

This energy is in fact twice the surface energy (γ) of the planes' material (equation [2.20]), so we can write:

$$2\gamma = \frac{A}{12\pi D_0^2} \quad [2.34]$$

which allows us to write:

$$\gamma = \frac{A}{24\pi D_0^2} \quad [2.35]$$

A number of experimental studies allow us to take D_0 = 0.165 nm [GEO 00, ISR 99]. Equation [2.35] gives a simple relationship between surface energy and Hamaker's constant. To determine Hamaker's constant, all that is required is knowledge of the surface energy of the material determined with a simple wettability test (see section 1.2.3).

Tribology 75

2.5.2. Surface forces measurements

Surface forces are primarily studied with two kinds of devices: the *surface forces apparatus* (SFA) or the *atomic force microscope* (AFM).

2.5.2.1. The surface forces apparatus (SFA)

The principle of the SFA is to carefully bring together a mobile surface (a plane or sphere) and a fixed surface (a plane) and to measure, during approach and withdrawal, their forces of interaction as well as the distance between them [GEO 94].

Figure 2.26 shows the evolution of the force depending on the phase of approach or withdrawal for two carbon surfaces, a sphere and a plane, in a dry atmosphere. During approach, a jump to contact resulting from Van der Waals forces occurred. During withdrawal, some hysteresis is noted and the maximum adhesive force (436 mN m^{-1}) can easily be deduced. Based on this quantity, the surface energy of carbon can be calculated by applying the JKR model (see Table 2.1) and equation [2.21]:

$$\gamma = \frac{F_{ad}}{3\pi R} \quad [2.36]$$

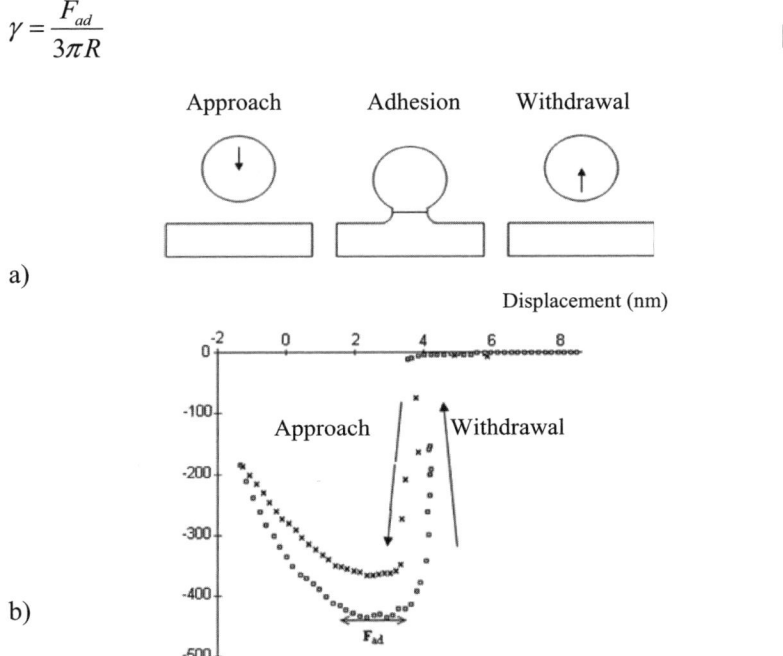

Figure 2.26. *Forces of adhesion of a sphere on a plane. Forces curve obtained for two carbon surfaces with an SFA used in dry air: a) displacement of the sphere; b) force curves [GEO 00]*

Figure 2.27 shows the impact of humidity on the force profile for two carbon surfaces. The curves shown are for the withdrawal phase. We see clearly that the maximum adhesive force (476 mN m^{-1}) measured in a wet atmosphere (relative humidity 30%) is greater than the equivalent for measurements in a dry atmosphere. Moreover, the shape of the curve clearly provides evidence for the generation of an interfacial meniscus (the withdrawal phase is accompanied by a strong tail on the curve).

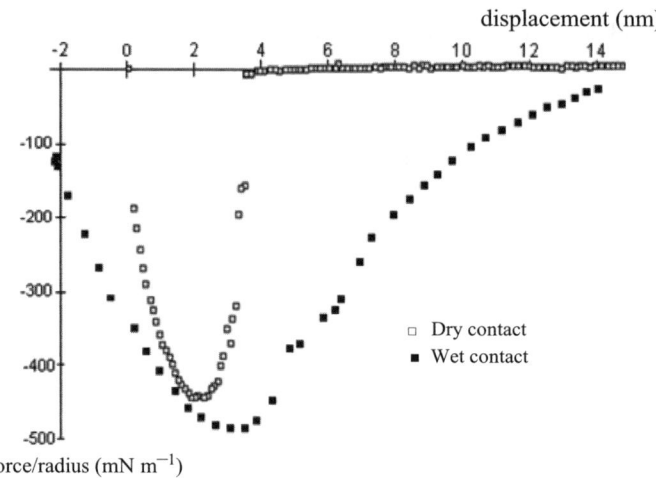

Figure 2.27. *Force profile relative to displacement during withdrawal of the sphere from the carbon plane; curves recorded in dry and wet conditions [GEO 00]*

2.5.2.2. The atomic force microscope (AFM)

The principle of the AFM was presented in section 1.2.2.3.3. This technique offers better resolution both in terms of the applied force as well as vertical displacement, and also allows improved spatial resolution as the radius of curvature of the AFM tip ranges from a few nanometers to a few tens of nanometers (whereas spheres used in SFAs have radii of several millimeters).

We now present some measurement results obtained from experiments carried out on immersed surfaces in different electrolytes.

Two metallic surfaces dipped in an electrolyte are brought close together. Depending on the nature of the environment, the materials, the adsorbed species and the pH of the solution, the surfaces may be subject to attraction or repulsion under the action of a range of forces such as hydrophobia, hydration, Van der Waals forces or electrostatic forces [BHU 05].

Consider using an AFM to study the particular case of the contact between different materials in a solution with differences in pH. We know that for a given electrolyte, the material's surface will develop a negative charge beyond a certain threshold pH (pH_s) and will develop a positive charge below this threshold. The point corresponding to a surface charge of zero (pH = pH_s) is known as the isoelectric point or IEP.

As the tip of the AFM is made of silicon nitride (Si_3N_4) (whose IEP corresponds to a pH of 6), then depending on the nature of the material (i.e. its IEP) brought into contact with the tip in a given electrolyte, the two surfaces will develop same-sign or opposite sign charges and repulse or attract.

Figure 2.28a shows force curves corresponding to the contact between the tip of the AFM (Si_3N_4) and a silicon oxide sample, in a 1 mM NaCl electrolyte solution, at pH 4 and at pH 8.5. Silicon oxide has an IEP of pH 2, so its interaction with silicon nitride (of IEP pH 6) will be attractive between pH 2 and pH 6 (the case for pH 4) and repulsive otherwise (the case for pH 8.5) [MARTI 95]. For the pH 4 case, we note that there is a large degree of hysteresis corresponding to strong adhesion between the surfaces. However, at pH 8.5 the force curve shows repulsion between the surfaces.

Figure 2.28b shows force curves for a polycrystalline nickel sample in contact with the tip of the AFM (Si_3N_4) in a 1 mM NaCl electrolyte solution at pH 3.3 and at pH 10.5. Nickel is negatively charged irrespective of pH whereas silicon nitride (IEP pH 6) is positively charged at pH 3.3 (attraction between Ni and Si_3N_4), and negatively charged at pH 10.5 (repulsion between Ni and Si_3N_4). The attraction, which is associated with a strong adhesive force, is clearly illustrated by the hysteresis in the adhesion measured at pH 3.3 [GAV 02a].

As a result of experiments in nanotribology, a new approach to the interpretation of friction has been introduced by Israelachvili [ISR 94]. He has shown that friction is not correlated to the force of adhesion (strength of formed junctions), but to adhesion hysteresis, i.e. the energy expended during the adhesion–rupture cycle.

This energy represents the difference between the energy required to establish and break the contact. It reflects the irreversibility of the force responsible for the adhesion, whereas in the Tabor and Bowden model, it is the value of this force that determines the degree of friction.

Figure 2.29 enables the unambiguous verification of this model. These results are from the system described above and are shown in Figure 2.28a. Several force curves were recorded at different pH levels and friction forces were measured with a lateral force microscope (LFM) (see section 1.2.2.3.3). We note perfect correlation between the adhesion hysteresis and the friction force.

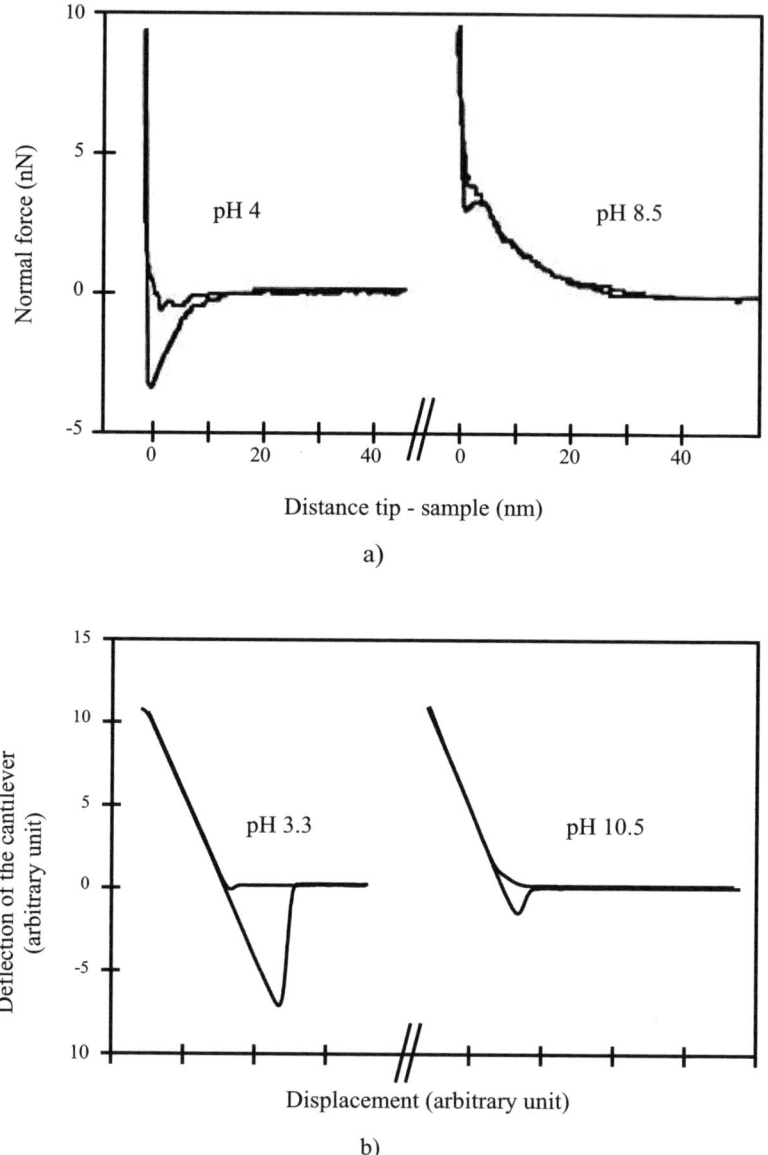

Figure 2.28. *a) Force curve recorded at pH 4 and pH 8.5 in a solution of 1 mM of NaCl between the silicon nitride tip of the AFM and a silicon oxide sample [MARTI 95]; b) force curve recorded at pH 3.3 and pH 10.5 in a solution of 1 mM of NaCl between the silicon nitride tip of the AFM and a nickel sample [GAV 02a] (see Figure 1.18 for the interpretation of force curves)*

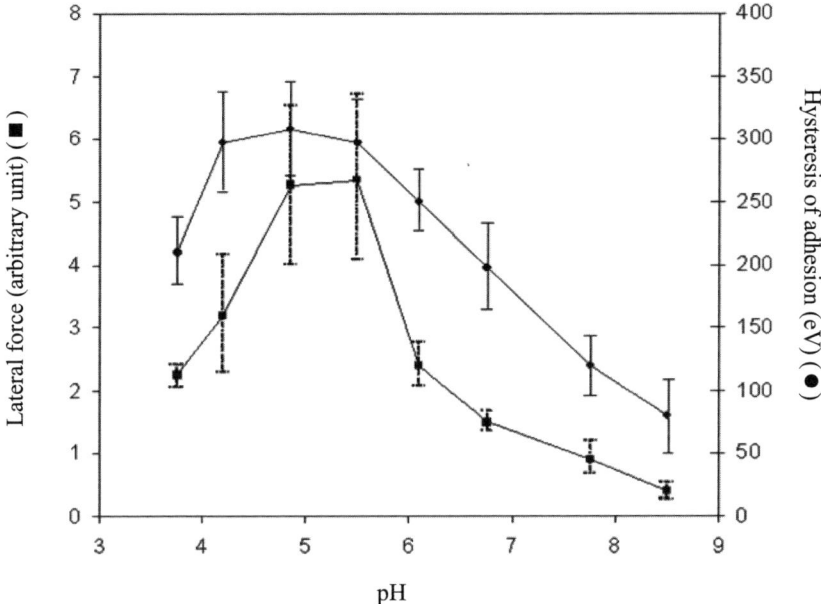

Figure 2.29. *Lateral force (friction force) (■) and expended energy (hysteresis of adhesion) (●) measured with AFM/LFM at different levels of pH in a solution of 1 mM of NaCl between the silicon nitride tip of the AFM and a silicon oxide sample [MARTI 95]*

2.5.2.3. Application: surface forces and micromanipulation

As shown in Table 2.3, as the dimensions of an object reduce, there is an increase in the surface area to volume ratio, which results in surface forces playing an increasingly important role in interactions. This is perfectly illustrated in the field of micromanipulation – the positioning and precise movement of micron-sized objects.

	Radius (m)	Surface/volume (m^{-1})
Atom	10^{-10}	3×10^{10}
Grain of sand	2×10^{-4}	15×10^3
Table-tennis ball	0.02	150
Soccer ball	0.1	30

Table 2.3. *Some values of the surface/volume ratio*

The drive towards device miniaturization and the need to produce components for micro-electro-mechanical systems (MEMS) motivates the need for grippers able to hold and move micron-sized components. However, as surface forces become greater on these scales, the force of adhesion (due to Van der Waals, electrostatic or capillary forces) renders the micromanipulation of these objects intrinsically difficult.

Although it is relatively easy to exploit surface forces to hold an object due to its adhesion to the grippers it is, on the other hand, difficult to release the object and thus free the grippers for a new manipulation. One solution to this problem consists of tilting the grippers by some angle θ in order to minimize the force of adhesion. This procedure is illustrated in Figure 2.30.

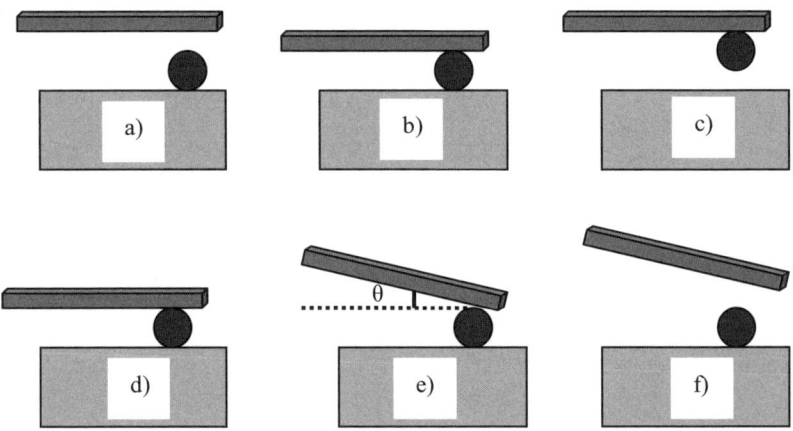

Figure 2.30. *Micromanipulation sequence: a) approach of the gripper; b) object–gripper contact; c) the object is taken hold of; d) the object is laid down; e) the gripper is tilted and the object is freed; f) the gripper is removed*

The gripper comes into contact with the object, adheres to it and lifts it. For this operation to succeed, the object needs to be "torn" from the substrate on which it is resting; in other words, the object–substrate bond must be broken. The adhesive force between the gripper and object must therefore be greater than the object–substrate bond. Once held by the gripper, the object must be set down in a different position but on the same substrate. For this to be possible, the object–substrate bond must be greater than the object–gripper bond. This can be achieved by tilting of the gripper according to an angle θ (with respect to the horizontal plane). This reduces the object–gripper adhesive force (which is multiplied by a factor cos θ), which is sufficient to make the object–substrate adhesion greater than that of the object–

gripper. For a detailed analysis (and calculations) of the pick and place, see [HALI 02] and [ROL 00].

Depending on the nature of the object to be moved, it is also possible to choose two surfaces made of materials A and B and a gripper made of material C in order to move the object from A to B without having to tilt the gripper by an angle θ. Careful selection of the three materials can ensure that the gripper–object adhesive force is greater than that between material A and the object, but less than the adhesion between material B and the object.

In order for the manipulation described above to be successful, the following double inequality between Hamaker's constants need to be satisfied [ROL 00]:

$$A_{(object-material\ B)} > A_{(gripper-object)} > A_{(object-material\ A)} \quad [2.37]$$

2.5.3. *Nanofriction*

Figure 2.31 shows the variation of the friction force as a function of the normal load for a silicon tip (of a few nanometers of radius of curvature) covered with an amorphous layer of carbon, sliding on the surface of mica or carbon samples.

Because of surface forces, we note the existence of non-zero friction even with no applied load (Ft* = 0 to 15 nN depending on the radius of curvature of the tip and the nature of the samples tested) [PIE 99, SCH 97a, SCH 97b]. The friction force cancels for a negative applied normal force P* (maximum adhesive force) between 2 and 50 nN.

This result illustrates the complex nature of nanotribology compared to classical tribology, and shows that problems linked to friction and wear in microsystems cannot be resolved through the naive transposition of laws and empirical solutions that are valid on the macroscopic scale. Indeed, the size factor (the scaling from macro to micro, or even nano) is not beneficial to miniaturization because, in microsystems, the forces dissipated by friction in mechanical links are considerable [MIN 98].

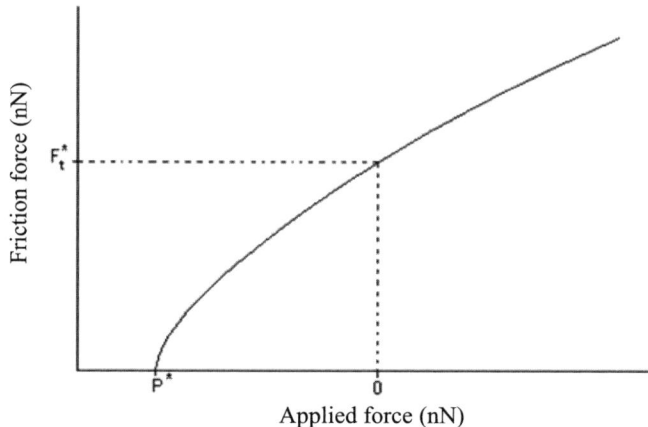

Figure 2.31. *Friction force relative to the normal load for a silicon tip covered with an amorphous film of carbon, and sliding on the surface of mica or carbon samples. F_t^* and P^* range from 0 to 15 nN and from 0 to 50 nN, respectively [PIE 99, SCH 97a, SCH 97b]*

2.6. Wear

Wear, corrosion and the specific degradation of polymers under ultraviolet light are the main modes of material degradation. Wear typically induces some degree of material loss. In the case of a hard material A in sliding contact with a soft surface B, we write the wear experienced by surface B as:

$$\frac{V}{L} = K \frac{F_n}{H} \qquad [2.38]$$

where V represents the total volume of material removed, L the total sliding distance, F_n the normal applied load and H the hardness of the softer material. K is a (dimensionless) wear coefficient which can take values ranging from 10^{-10} (mild wear) to 10^{-3} (severe wear).

The coefficient K is sometimes replaced by a coefficient K', which is given by:

$$K' = \frac{V}{LF_n} \qquad [2.39]$$

expressed in $mm^3 \, mN^{-1}$.

Several methods can be used to quantify wear, such as:
– weighing of samples once the tests have been performed;

– quantifying the volume of removed material through the use of 3D tactile or optical profilometers;

– filtration and analysis of the oils and wear debris found in lubricants;

– surface activation which consists of marking it with radiotracer isotopes and monitoring the wear through analysis of the radioactive signal emitted; provided it has been properly calibrated, the intensity of this signal can yield the depth of wear.

2.6.1. *The different forms of wear*

It is usual to classify wear in terms of four different categories: adhesive wear, abrasive wear, fatigue wear and tribochemical wear (see Figure 2.32).

2.6.1.1. *Adhesive wear*

Adhesive wear (illustrated in Figure 2.32a) is characterized by the appearance of junctions (or microwelds) between the surfaces that are subject to friction. When these junctions are weak, shear occurs at the interface of the two surfaces and there is no wear. However, when junctions are strong, the softer material is subject to shearing and, as a consequence, is transferred onto the harder material.

2.6.1.2. *Abrasive wear*

Abrasive wear (illustrated in Figure 2.32b) occurs when a hard material is put into contact with a soft material. This type of wear can cause scratches, wear grooves and lead to material removal.

2.6.1.3. *Fatigue wear*

Surface fatigue wear (illustrated in Figure 2.32c) occurs when a material is subject to cyclical stresses. Due to strains introduced in the superficial layers of the material, cracks that are parallel to the surface develop within the material. When they reach a critical size, they generate flake-like debris. This phenomenon is also referred to as delamination wear.

2.6.1.4. *Tribochemical wear*

Tribochemical wear (illustrated in Figure 2.32d) is a phenomenon which involves the growth of a film of reaction products due to chemical interactions between the surfaces in contact with each other and the surrounding environment.

One of the most common forms of tribochemical wear is tribo-oxidation wear. The increase in temperature due to friction accelerates the growth of an oxide film which detaches from the surface when it reaches a certain critical thickness. The

84   Materials and Surface Engineering in Tribology

debris thus generated can take part in the wear process or be evacuated from the friction path.

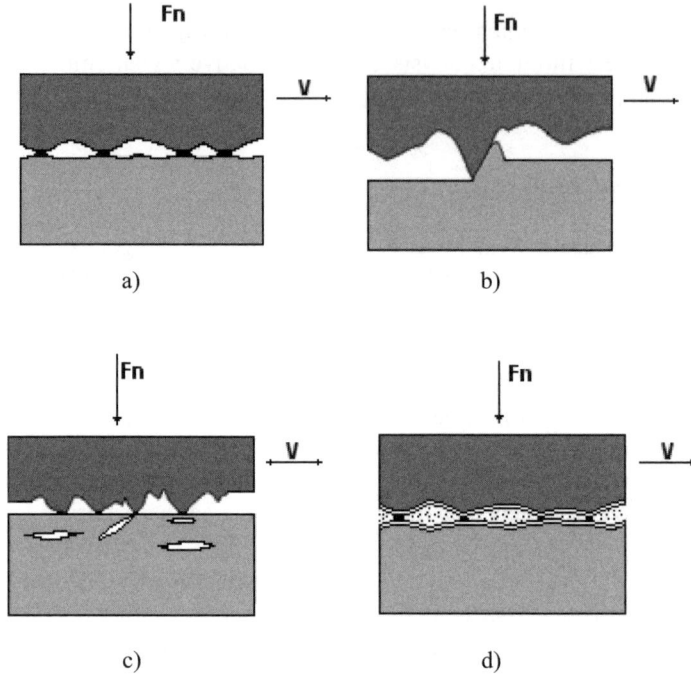

Figure 2.32. *Main forms of wear: a) adhesive wear; b) abrasive wear; c) fatigue wear; d) tribochemical wear*

2.6.2. *Wear maps*

Wear maps are used to represent the different types of wear as a function of tribological parameters [LIM 97]. Figure 2.33 shows the wear map of a steel/steel pair in a pin-on- disk-type contact. The different domains are plotted as a function of normalized parameters.

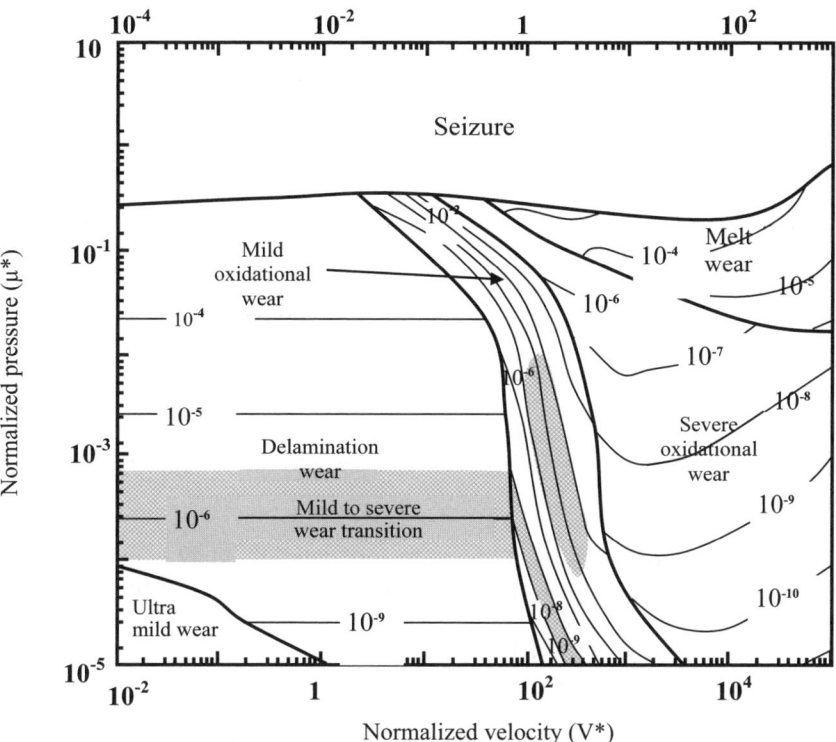

Figure 2.33. *Wear map of steel including the different forms of wear as well as corresponding values for the rates of wear [LIM 87]*

The normalized speed V* is defined by:

$$V^* = v\frac{a}{k} \qquad [2.40]$$

where v is the sliding speed, a is the radius of the contact area and k is the thermal diffusion coefficient of the metal.

The normalized pressure P* is given by:

$$P^* = \frac{F_n}{A_n}H \qquad [2.41]$$

where F_n is the applied load, A_n the apparent contact surface area and H the hardness.

The normalized rate of wear is:

$$W^* = \frac{W}{A_n} \qquad [2.42]$$

where W is the rate of wear expressed in terms of the volume of removed matter per unit of applied load and per unit of sliding distance and A_n is the apparent contact surface area.

2.6.3. *Interface tribology: third body concept*

Interface tribology distinguishes first bodies (those involved in the friction) from the third body which comes in between and partially or completely separates this material couple.

In dry (lubricant-free) friction, the third (or "natural") body is made up of wear debris forming a layer which can accommodate velocity differences between the first two bodies. This layer ensures load-carrying capacity, protects the surfaces and can act as a lubricant, as has been demonstrated in the case of silicon carbide couples [TAK 94a, TAK 94b].

Liquid and solid lubricants are considered as artificial third bodies as they are introduced into the contact area by external means.

The notion of the tribological circuit was introduced to describe the evolution of the third body in the contact area [BERTH 88] (see Figure 2.34). Initially, wear particles are removed from the surfaces subjected to friction and come to provide an internal source. If a solid lubricant (or external source) is used, the third body will comprise torn particles and lubricant. As new wear particles are removed from the antagonistic surfaces and enter the contact area, part of the third body particles are evacuated from the surface.

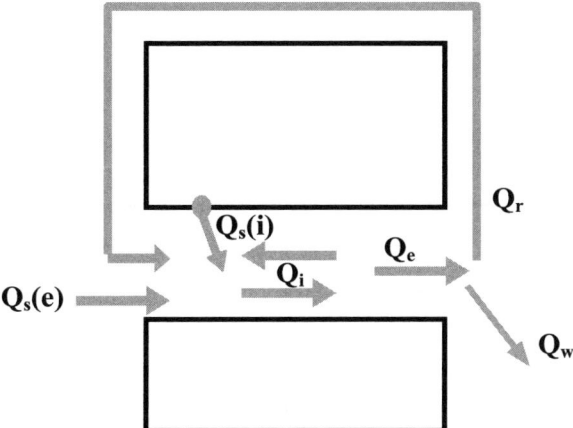

Figure 2.34. *Principle of the tribological circuit and evolution of the third body [BERTH 88] (Q_e: flow of the third body escaping from the contact; Q_w: wear flow; Q_r: recirculation flow; $Q_s(i)$: internal source flow, representing the particles detached from the first bodies; $Q_s(e)$: external source flow, due to the artificial third body introduced in the contact; Q_i: internal flow, or flow of the third body circulating between the first bodies)*

The third body can also be defined as an operator that separates the first bodies and transmits the load (bearing) between the first bodies, while accommodating the greater part of their difference in speed [BERTH 98].

As shown in Figure 2.35, velocity accommodation between the first bodies can occur in five different sites, i.e. the two first bodies, the third body and the two intermediate layers between the two first bodies and the third body [BERTH 88, BERTH 92]. Moreover, within each site, four modes of velocity accommodation can occur: elastic, rupture, shear and roll. There are twenty velocity accommodation mechanisms (5 sites × 4 modes) and, depending on the nature of the materials and the conditions, one or more of these are activated. In fact, in the course of friction, one particular mechanism can be stopped and one or several other mechanisms can be triggered.

Berthier provides an overview of third body theory and its uses in solving problems involving friction and wear [BERTH 05].

88 Materials and Surface Engineering in Tribology

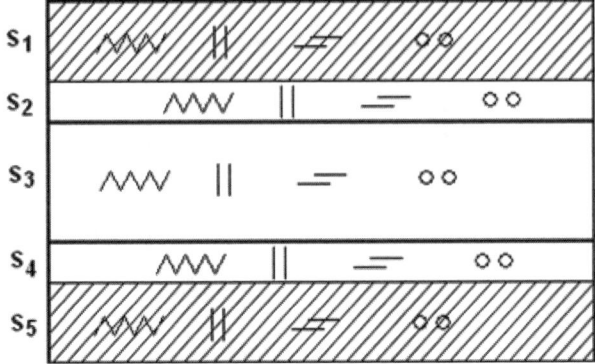

Figure 2.35. *The different mechanisms of velocity accommodation [BERTH 88, BERTH 92]:
S1, S5 = 1st bodies; S2, S4 = screens; S3 = 3rd body; M1 /\/\/\ = elastic mode;
M2 || = rupture mode; M3 ⇌ = shear mode; M4 ○ ○ = roll mode*

2.6.4. The PV product

Even although we can currently identify and classify many different forms of wear (abrasion, adhesion and surface fatigue), practical experience shows that wear is often due to several mechanisms occurring simultaneously. This makes it difficult to reliably predict the behavior of tribological systems. This complexity is mainly due to the following six factors:

– contact pressure varies with wear as the true contact area changes;

– the chemical composition, texture and mechanical characteristics of a surface are modified during frictional contact;

– the exact transformation undergone by the third body and wear debris, as well as the part they play in the wear mechanism itself, are not yet fully understood;

– the results of laboratory tests can be influenced by the mechanical characteristics of experimental instruments;

– the chemical interaction of lubricants (in particular, certain additives) with the surfaces subject to friction and the consequences for the tribocorrosion (and wear) of materials are problems not yet fully understood; and

– the dissipation of energy during friction is a problem not yet resolved.

From the practical point of view, the performance of a material can be described by the maximum value of the PV product (pressure × speed) which it can withstand. In the case of polymer–steel friction, values ranging between 0.06 MPa ms^{-1} for PTFE and 3.5 for polyimide have been reported. In car manufacturing, the PV product can range from 60 to 250 MPa ms^{-1} for a standard crankshaft (from 200 to

250 MPa ms^{-1} for a big end), and it can be as high as 650 MPa ms^{-1} for racing engines (1000 MPa ms^{-1} for a big end) [GEO 00, LIG 04].

2.7. Lubrication

The purpose of lubrication is to extend the lifespan of those parts subject to friction, as well as that of the mechanisms and machine parts in which they are involved.

Three types of lubricants are used:
– oils;
– greases; and
– anti-friction materials.

2.7.1. *Oils [AYE 92, MANG 01, RUD 03]*

2.7.1.1. *The notion of viscosity*

Lubrication oil composition is complex because, as well as containing standard oil, they comprise a great many additives such as anti-oxidants, anti-rust products, detergents, anti-foaming agents or extreme-pressure additives.

The principal characteristic of oil is its viscosity, usually noted η. This quantity describes the fluid's resistance to flow. In the SI system, it is expressed as Pa s and in the CGS system, the unit is the Poise (1 Pa s = 10 Poises). For example, the viscosity of water is 1 mPa s at 20°C while that of a car engine lubrication oil is 10 mPa s and that of a molten polymer is of the order of 1000 Pa s.

In place of the dynamic viscosity, we sometimes use the kinematic viscosity (v) which is defined by:

$$v = \frac{\eta}{\rho} \qquad [2.43]$$

where ρ is the volumetric mass of the liquid. v is given in m^2 s^{-1}, but the Stokes (St) where 1 cSt = 10^{-6} m^2 s^{-1} is also often used.

In order to facilitate the ease of flow and contact lubrication, oil must retain its viscosity at high temperatures yet remain fluid at low temperatures. This is usually achieved through the use of specific additives, such as polymethacrylates, which guarantee oil fluidity in cold weather while allowing it to remain viscous enough to prevent metal–metal contact at high temperatures.

2.7.1.2. *The viscosity index and the SAE standard*

The viscosity index or VI accounts for variations in oil viscosity as a function of temperature. If oil viscosity varies greatly with temperature, its viscosity index is low; conversely, a high viscosity index characterizes oils that are relatively impervious to temperature changes. This is determined relative to two reference oils to which are assigned indices 0 and 100. The viscosity of the oil under study is measured at 40°C and 100°C and the values obtained are then compared with those of the reference oils.

VI is calculated using the following equation:

$$VI = 100 \frac{L-U}{L-H} \qquad [2.44]$$

where U is the viscosity of the oil under study and where L and H are the viscosities of the reference oils with VI = 0 and VI = 100, respectively. All three oils must have identical viscosity at 100°C.

To classify oils into categories, the American *Society of Automotive Engineers* has established the SAE standard which is used universally. It defines 11 levels or "grades" characterized by values of viscosity situated within an interval defined by the standard. These levels of viscosity are measured in cold temperature conditions (–5 to –40°C) for winter grades (W) and in hot conditions (100°C) for the summer grades.

Oils satisfying the viscosity limits for either hot or cold conditions are called monograde. Oils meeting the viscosity requirements for both hot and cold conditions are called multigrade.

Examples of monograde oils include the following:

– SAE 10W: smaller numbers are associated with oils that retain high fluidity even in cold conditions. The number corresponds to one of the six grades: 0, 5, 10, 15, 20 or 25.

– SAE 40: higher numbers are associated with oils that retain high viscosity even at high temperatures. The number corresponds to one of the five grades: 20, 30, 40, 50 or 60.

Examples of multigrade oils include:
– SAE 5W20; and
– SAE 10W40.

The numbers given before and after the letter W have the same meaning as those described above for monograde oils.

2.7.1.3. *The Stribeck curve*

To illustrate the behavior of a lubricant in action, we consider in Figure 2.36 the case of a plane bearing and monitor the evolution of the friction coefficient when it is in use. The variation of this coefficient is plotted as a function of the Sommerfeld number $S = \eta V/P$, where η is the oil's dynamic viscosity, V is the relative velocity of the opposing surfaces and P is the normal load.

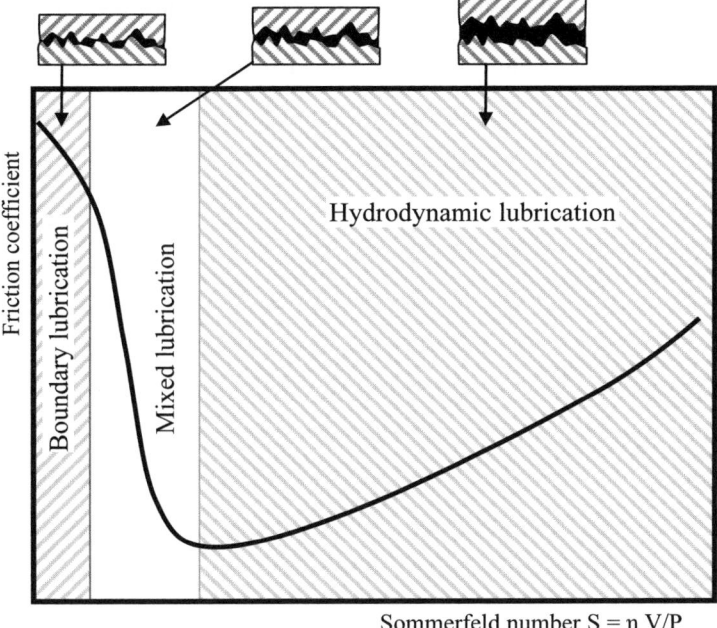

Figure 2.36. *The Stribeck curve*

At the start we observe a high friction coefficient due to the metal–metal contact. In this *lubrication limit* zone, the mean thickness of the lubricating film is lower than the mean roughness height of the opposing surfaces.

The second phase is called *mixed lubrication* (see Figure 2.36) and corresponds to metal–metal friction in the presence of a discontinuous lubricating film. When the lubricating film completely separates the surfaces, the friction coefficient passes through a minimum value and then begins to increase due to viscous friction within the lubricant. From then on, it is the rheological properties of the lubricant and its adhesiveness to the surfaces in contact that determine the tribological behavior of the system. This domain corresponds to the regime of hydrodynamic lubrication (with a continuous lubricating film) and corresponds to the optimum operating zone of the plane bearing.

2.7.1.4. *The different types of oils*

2.7.1.4.1. Mineral oils

These are petroleum-based oils containing several kinds of hydrocarbons whose nature depends on the particular crude oil used. The main hydrocarbon families found in mineral-based oils are paraffinic, naphthenic and aromatic oils (see Table 2.4), the latter being the dominant component accounting for more than 70% of the weight of mineral-based oils.

Name	Chemical formula
Structure of linear (n-paraffin) paraffinic-type hydrocarbon	$CH_3-(CH_2)_n-CH_3$
Structure of aromatic-type hydrocarbons	
Structure of olefinic-type hydrocarbons	$CH_3-(CH_2)_n-CH_2-CH=CH-CH_2-(CH_2)_n-CH_3$
Structure of naphthenic-type hydrocarbons	

Table 2.4. *The main types of hydrocarbons found in mineral oils*

Apart from these primary constituents, mineral oils also contain oxygenated, nitrogenated and sulfurated components that can make up as much as 3% of the total product weight.

Finally, we note that mineral oils have a generally poorer lubrication performance than compared to synthetic oils and that they are also more environmentally hazardous.

2.7.1.4.2. Synthetic oils

Synthetic oils can be classified into four main categories based on their chemical nature: polyolefins, polyglycols, silicones and esters (see Table 2.5).

Polyolefins are largely used in engines and gear lubrication, whereas polyglycol-based oils are used as brake fluids or as lubricants to prevent metal wear. Silicone-based oils are noted for their very high viscosity level and can be used, for example, as a damping liquid or sliding lubricant. Ester-based oils provide excellent resistance to oxidation and good thermal stability. They are therefore widely used in high-temperature systems.

Name	Chemical formula
Aliphatic ester	$R'-(CH_2)_n-\underset{\underset{O}{\|}}{C}-O-R$
Silicone	$CH_3-\underset{\underset{CH_3}{\|}}{\overset{\overset{CH_3}{\|}}{Si}}-O-\left[\underset{\underset{CH_3}{\|}}{\overset{\overset{CH_3}{\|}}{Si}}-O\right]_n-\underset{\underset{CH_3}{\|}}{\overset{\overset{CH_3}{\|}}{Si}}-CH_3$
Polyglycols	$HO-(CH_2-\underset{\underset{CH_3}{\|}}{CH}-O)_n-H$
Polyalphaolefins (PAO)	$\left[-\underset{\underset{R}{\|}}{\overset{\overset{CH_3}{\|}}{C}}-CH_2-\right]_n$

Table 2.5. *The different types of synthetic oils*

2.7.1.4.3. Vegetable oils

Vegetable oils are intrinsically biodegradable and non-toxic; sunflower, soya and rapeseed oils can be as efficient as some mineral oils at low temperatures. The development of new antioxidant additives has now overcome the main drawback of these lubricants – their limited resistance to oxidation – and many studies have shown that these oils can now successfully rival mineral oils of similar viscosity [KAB 95, MAS 99]. There are strong arguments for the increased use of vegetable oils as environmental standards become stricter and as costs of petroleum-based products increase, particularly considering the opportunities to cultivate fallow land.

2.7.1.4.4. Additives

Lubrication oils are systematically used together with a range of additives which improve their rheological and chemical properties. These are grouped in two classes:

– additives acting on the properties of the lubricant such as anti-foaming, antioxidant agents and viscosity index improvers; and

– additives interacting with the opposing surface to form a protective film (i.e. anti-wear and extreme-pressure additives, corrosion inhibitors, detergents and dispersing agents).

2.7.1.5. *Greases*

Greases consist of basic mineral or synthetic oil that is mixed with a thickener containing an alkaline or alkaline earth element such as lithium, calcium, potassium, sodium or barium. The properties of a type of grease and its performance at high temperature greatly depend on the nature of its constituents. In contrast to oils, greases are advantageous in that they do not flow and remain in the contact areas during friction.

2.7.1.6. *Anti-friction materials*

Anti-friction materials are used in those cases where oils or greases are unsuitable. They must meet three essential requirements:

– be soft and easy to shear;

– be perfectly adherent to the surfaces of the materials to be protected; and

– be stable and impervious to chemical or structural changes in the course of their application.

These materials can be divided into four main categories:

– lamellar solids;

– sliding varnishes;

– soft metals; and

– fluorides, sulfides and oxides.

2.7.1.6.1. Lamellar solids

Graphite and molybdenum disulfide (MoS_2) are the two most widely used. They are characterized by a structure of fine layers known as a lamellar structure (see Figures 2.37 and 2.38). This results from the stacking of atomic planes that are loosely bound by weak bonds. During friction, when shearing occurs parallel to these planes, they slide along one another and thus act as a lubricant in allowing the relative sliding of the solids in contact.

These materials are generally used as thin films (ranging from a few nanometers to a few tens of nanometers) physically deposited on the metallic surfaces expected to undergo friction. This is usually achieved by the sputtering of a graphite or molybdenum disulfide target onto the metallic surfaces undergoing friction. They are also used as additives in composite materials designed for tribological applications.

These two materials are very sensitive to their environment. Graphite loses its lubricating properties in a dry atmosphere, whereas in humid conditions the layer of water adsorbed onto its surface grants it excellent lubricating properties. In contrast, for molybdenum disulfide, friction and wear noticeably increase significantly with increased humidity [FLE 97].

The tribological properties of graphite and molybdenum disulfide can be improved through the use of certain additives such as PTFE. Thanks to its hydrophobic nature, this additive results in improved properties for the MoS_2-PTFE composite in a humid atmosphere.

The lubricating properties of graphite can also be improved through the addition of fluoride, which will induce the generation of $(CFx)_n$-type composition, where fluoride atoms replace carbon atoms without modifying the lamellar structure of the graphite. Because fluoride noticeably weakens the bonds between atomic planes, it reduces resistance to shear which, in turn, greatly reduces the friction coefficient [PAU 96].

While the friction and wear of graphite are greater in vacuum than in air, molybdenum disulfide has a low friction coefficient in a vacuum (0.15 in air; 0.02 in a vacuum) even at high temperatures. In vacuum it remains stable to 1000°C, whereas in air it oxidizes readily at 400°C yielding MoO_3 and SO_2. These properties make MoS_2 particularly well-suited for spatial applications.

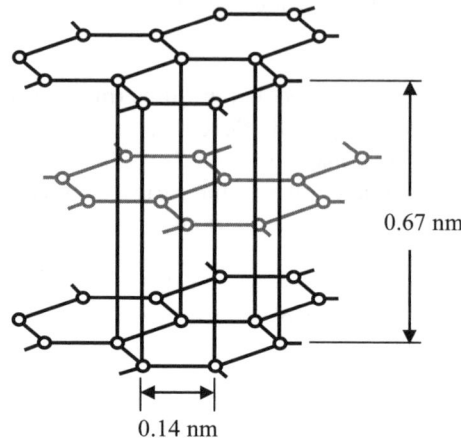

Figure 2.37. *Crystallographic structure of graphite*

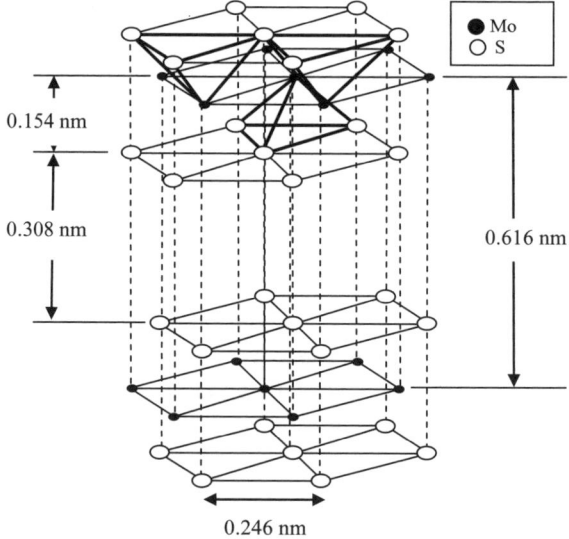

Figure 2.38. *Crystallographic structure of molybdenum disulfide (MoS_2)*

2.7.1.6.2. Varnishes

These are composite materials consisting of solid lubricant particles (usually MoS_2, graphite or PTFE) dispersed into an organic or inorganic binding agent in the presence of a solvent. They are applied as a thin layer (ranging from a few microns to a few tens of microns) through spraying or dipping of previously degreased surfaces. They can be applied equally well to metals and to flexible materials such as rubber. Some of these coatings may be air-dried at room temperature, but others need to be oven-dried with a drying phase ranging from a few minutes to several hours.

2.7.1.6.3. Soft metal coatings

These are usually gold, silver, indium, lead and tin thin films that are either deposited chemically, electrochemically, or using CVD (Chemical Vapor Deposition) or PVD (Physical Vapor Deposition) techniques. Soft metal coatings are interesting as they can withstand loads and working temperatures far greater than sliding varnishes.

Studies carried on indium [BOWD 64], lead [HALL 86, PAU 96] or gold films [ANT 88, SPA 81] deposited onto metallic substrates have clearly shown that there exists an optimal film thickness (generally ranging between 0.1 and 1 micron) that minimizes the friction coefficient.

Figure 2.39 shows the typical variation of the friction coefficient as a function of the film thickness for two (ball-on- disk) steel surfaces where the disk is coated with a soft metal film.

This curve can be interpreted as follows: when the film is very thin, some surface asperities are not covered. As a result, the ball comes into direct contact with the substrate, yielding a high friction coefficient and rapid surface wear. Conversely, when the film is very thick, its ductility causes the ball to penetrate it thus increasing the contact area and the friction coefficient. In this case, the load is principally carried by the film.

The optimum thickness will therefore be that which allows integral coverage of the substrate by a continuous film sufficiently thin that the load is principally supported by the hard substrate.

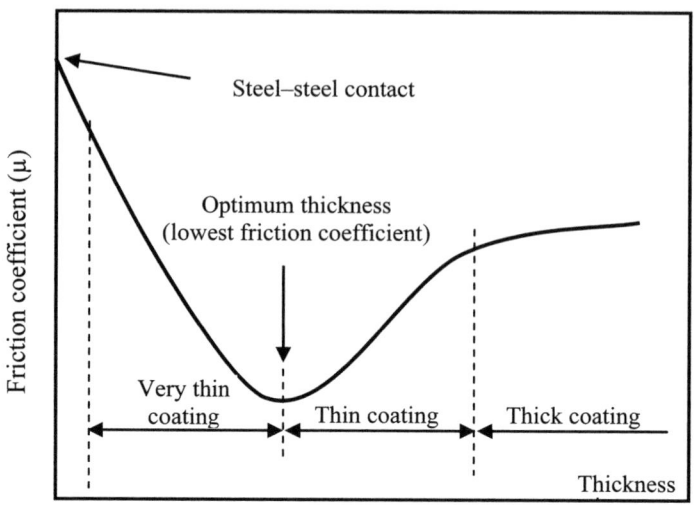

Figure 2.39. *Friction coefficient of two steel surfaces as a function of the soft metal coating thickness; optimum thickness usually ranges between 0.1 and 1 μm [ANT 88, BOWD 64, HAL 86, PAU 96, SPA 81]*

2.7.1.6.4. Fluorides, sulfides and oxides

PbO, SiO_2, B_2O_3, CaF_2, BaF_2 and PbS are the most common compounds used. Unlike most solid lubricants such as graphite and molybdenum disulfide or sliding varnishes, these materials present high thermal resilience. This means that instead of decomposing or losing their lubricating properties above certain temperatures (usually between 350°C and 500°C), they possess high thermal stability and can withstand higher temperatures (ranging between 600°C and 1000°C). They are generally applied through burnishing (the application of films of a few tens of microns) and can be used either as pure compounds or as mixtures of the type PbO-SiO_2, CaF_2-BaF_2 or B_2O_3-PbS. They can also be used in the composition of certain sliding varnishes or certain metal alloys deposited through the plasma spraying deposition technique.

2.8. Wear-corrosion: tribocorrosion and erosion-corrosion

Wear-corrosion refers to a specific process of material degradation which occurs when a material is placed in a corrosive environment and is subject to friction or erosion. In the first case (friction) we refer to the process as tribocorrosion and in the second case (erosion) we refer to the process as erosion-corrosion.

Under these conditions, the material degradation mechanisms involve mechanical processes (such as abrasion, yield and fracture) as well as chemical processes (such as passivating, dissolution and oxidation). The combination of these mechanical and chemical (or *chemo-mechanical*) effects can lead to the catastrophic acceleration of material degradation via extensive material loss. Indeed, the effect of synergy between mechanical wear and chemical corrosion results in a total volume of removed material (Vt) which can in fact exceed the sum of material separately removed through wear and corrosion.

The volume Vt is given as a function of three components:

$$Vt = Vw + Vc + Vs \qquad [2.45]$$

where Vw and Vc are the volume of material removed separately by the effects of wear and corrosion, respectively, and Vs represents the synergistic effect between wear and corrosion which can account for 20–70% of the total volume of material removed [GROG 92, MIY 90, MOON 91, ZHANGT 94].

2.8.1. *Tribocorrosion*

Tribocorrosion phenomena are observed in a large number of applications and in many different environments. Some typical examples of such occurrences include:

– in the moving parts of an engine such as pistons, cylinders and valves with lubricants;

– with eyeglass frames, due to friction with the skin in the presence of perspiration;

– with electrical connectors (when there is insertion and removal in a humid or corrosive environment);

– with joint prostheses (when friction occurs in a physiological liquid);

– with components used in plumbing and pump technology.

Tribocorrosion is also an important aspect of chemo-mechanical polishing (CMP) processes that affect the manufacturing of parts for micromechanics as well as silicon wafers for microelectronics and nanotechnology.

Erosion-corrosion occurs in pipelines used to transport corrosive liquids mixed with abrasive particles, such as an acid containing ceramic particles or seawater or petrol containing sand. These examples mainly concern the chemical and petrochemical industries, but erosion-corrosion attacks the majority of mechanisms operating in environments where sand is present in significant quantities.

Tribocorrosion studies are carried out through the use of specific set-ups comprising an electrochemical cell and a tribometer, allowing the sample to be subjected to friction under either a free or an applied electrochemical potential. The main quantities measured are the coefficient of friction and the corrosion current.

Figure 2.40 shows the principle of an experimental set-up designed for the study of tribocorrosion. When the sample is polarized, its behavior under friction is significantly modified.

Many studies have shown the strong influence of the electrochemical potential of a surface on its tribological behavior [MIS 93, TAK 96, YON 06].

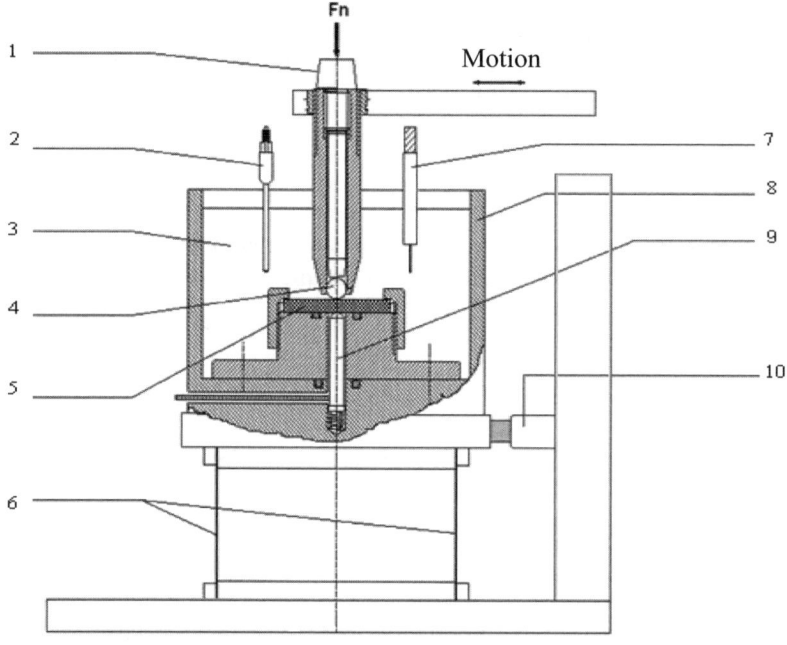

Figure 2.40. *Tribocorrosion analysis set-up: 1) normal applied load; 2) reference electrode; 3) electrolyte; 4) ceramic sphere; 5) sample; 6) spring plates; 7) platinum counter-electrode; 8) PTFE electrochemical cell; 9) towards the potentiostat; 10) force sensor (adapted from [TAK 96])*

The tribocorrosion of nickel and iron in a sulfuric environment (H_2SO_4 1N) is the focus of a previous study [TAK 96]. The electrochemical behavior of nickel in this environment is illustrated by the polarization curve as seen in Figure 2.41.

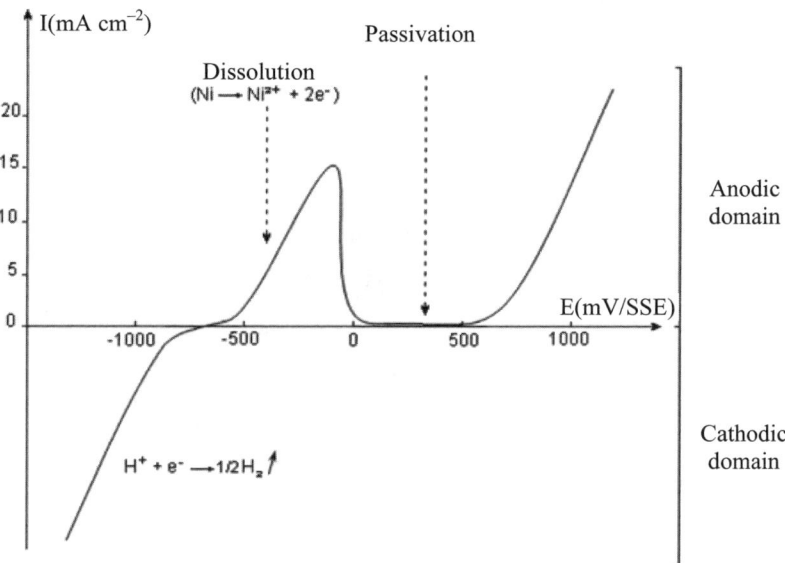

Figure 2.41. *Polarization curve (intensity-contact potential) of nickel in a sulfuric environment (H_2SO_4 1N); contact potentials are given relative to a saturated sulfate reference electrode (SSE)*

In the anodic domain, the curve shows a dissolution peak followed by a passivation plateau that corresponds to the formation of a film made up of a nickel oxide and nickel hydroxide mixture.

Several friction tests were performed on the material using an aluminum sphere of 5 mm in diameter and a normal load of 3.5 N. A number of different polarization potentials were used, corresponding to the cathodic domain (–1300 mV/SSE), the dissolution domain (–300 mV/SSE), the corrosion potential (–650 mV/SSE) and the passivation plateau (500 mV/SSE).

For the samples subjected to the corrosion potential or placed in the cathodic domain, we note only slight surface wear (see Figure 2.42c). The sample in the dissolution domain suffers significant losses of material due to the combined effects of corrosion and friction wear (see Figure 2.42a). Indeed, the synergy between electrochemical corrosion and wear leads to a proportional rate of material loss which increases as a function of the polarization potential and can represent up to 40% of the total loss of material (see Figure 2.43). Figure 2.42b shows that no wear appears on the surface of the sample polarized to 500 mV/SSE in the passivation domain, and we conclude that the passivation film provides excellent protection to the material.

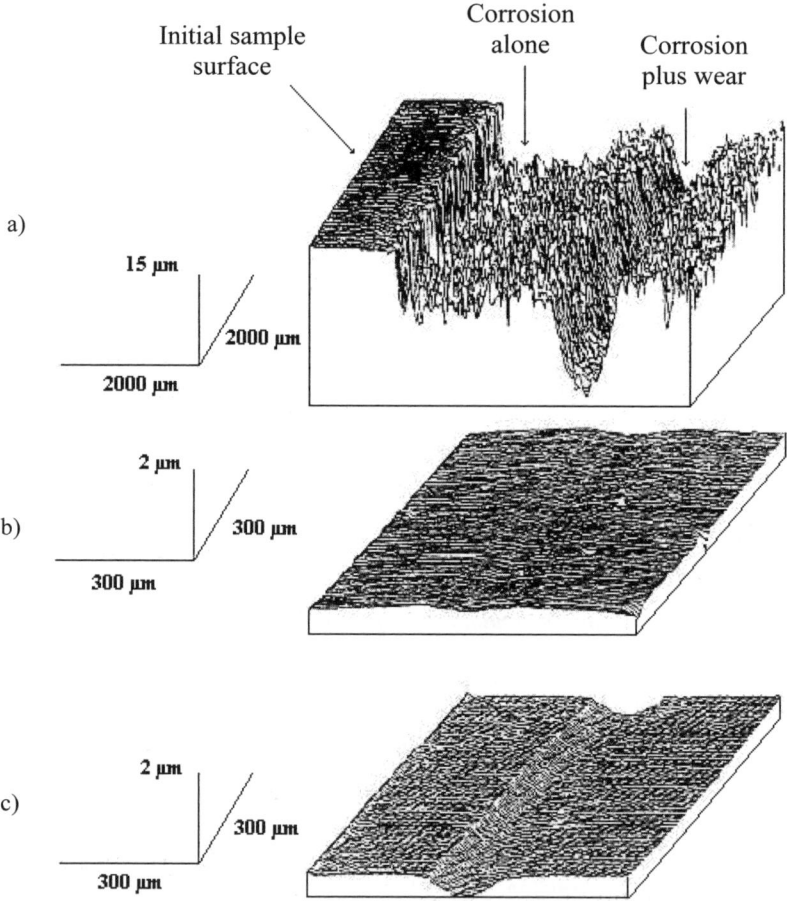

Figure 2.42. *Surface states of nickel samples following a friction test in a H_2SO_4 1N medium. The sample potential is: a) –350 mV, b) 500 mV, c) –650 mV (equally at –1300 mV). The reference surface (Figure 2.42a) is the initial sample surface protected during the tribocorrosion test [TAK 97c]. Potentials were measured relative to a saturated sulfate electrode*

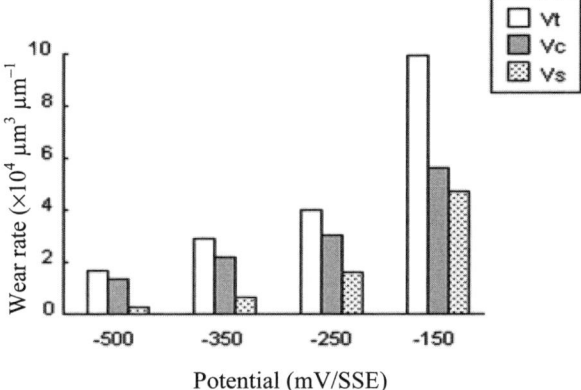

Figure 2.43. *Volumes of material removed by tribocorrosion as a function of the applied potential imposed on nickel. Vw, Vc, Vs and Vt are the volumes defined in the text (see equation [2.45]). When measured in the cathodic domain, volume Vw (–1300 mV) is 130 µm³ µm⁻¹) [TAK 97c]*

Other materials have a similar behavior to nickel. This is the case for iron when it is in a high concentration acidic medium ($H_2SO_4 > 10$ M) [MIS 93], and for copper [YON 06] or brass [QIU 02] when placed in a sodium laurylsulfonate medium with the addition of 0.1 M sodium sulfate.

For the case of brass in sliding contact against a silicon nitride sphere, Figure 2.44 shows the evolution of the friction coefficient as a function of the applied potential. When the potential is close to or greater than the corrosion potential (–0.444 V) measured relative to a saturated calomel electrode (SCE), a passivation film forms on the surface of the material. This acts as a lubricant and significantly reduces the friction coefficient as well as the surface wear (the measured friction coefficient is 0.07). Conversely, when the material is brought into the cathodic domain, the protective oxide film disappears due to hydrogen evolution which cleanses the surface and makes it more active. This yields a higher friction coefficient and a lower resistance to wear (the measured friction coefficient is 0.27).

However, note that the passivation film does not necessarily always grant the material good protection against wear. Indeed, the opposite is sometimes observed [BIE 00]. In this work, the authors studied the behavior under friction of tungsten against an aluminum sphere in media 0.5 M H_2SO_4 and 0.1 M $K_3Fe(CN)_6$ at pH 3. The applied load was 5 N and the sliding speed 62 mm s⁻¹.

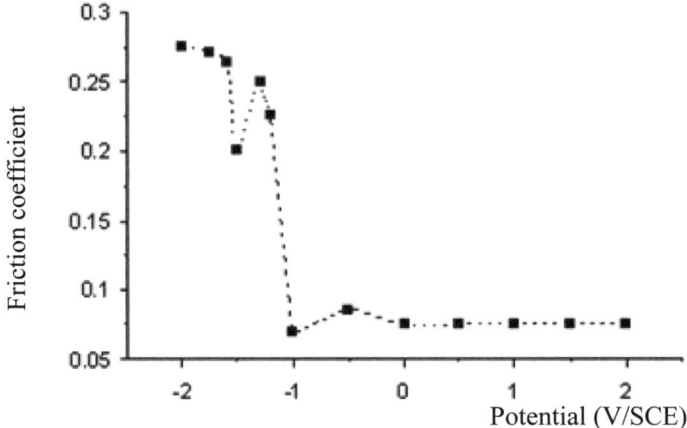

Figure 2.44. *Variation of the friction coefficient of brass against a SiO_2 sphere under a load of 9.8 N in a sodium laurylsulfonate solution with 0.1 M sodium sulfate. Values for the friction coefficient were recorded under several applied potentials [QIU 02]*

Figure 2.45 shows the wear volumes measured under different applied potentials. In an H_2SO_4 medium and under a cathodic potential, the detected wear is negligible. When no potential is imposed (i.e. with a corrosion potential of 60 mV), some material wear is measured as shown in Figure 2.45a. When a potential corresponding to the passivation of the surface (formation of tungsten oxide) is applied (380 mV relative to an Ag/AgCl electrode), a significant increase in wear is observed (see Figure 2.45b).

An additional test, carried out in a highly oxidizing medium ($K_3Fe(CN)_6$) with a corrosion potential of 380 mV relative to an Ag/AgCl electrode, recorded significant wear comparable to that obtained in a H_2SO_4 medium polarized in the anodic domain.

These results clearly show that oxidation of a surface contributes to its wear. The oxide degrades under friction and exposes the naked surface of the material which in turn oxidizes again when in contact with the electrolyte.

Similar results have been obtained for TiN coatings deposited on a steel substrate and subject to friction against an aluminum sphere in a borate medium (0.3 M H_3BO_3 and 0.075 M $Na_2B_4O_7$). When the material is polarized in the anodic domain, the oxidation of TiN and the generation of TiO_2 accelerate the surface wear under friction [BAR 01]. This is essentially due to the mechanical properties of TiO_2 which are much weaker than those of the TiN compound.

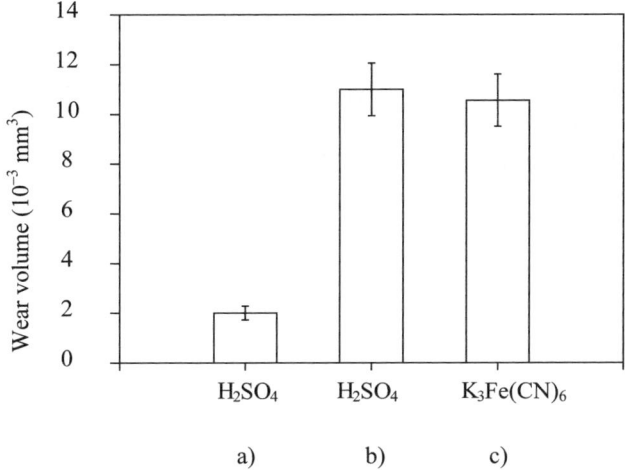

Figure 2.45. *Influence of the imposed electrochemical potential on tungsten wear: a) in H_2SO_4 0.5 M (under free potential E = 60 mV); b) in H_2SO_4 0.5 M (under imposed potential E = 380 mV); c) in $K_3Fe(CN)_6$ (under free potential E = 380 mV). The potentials are given relative to an Ag/AgCl electrode. Friction was imposed with an aluminum sphere of 5 mm in diameter sliding alternatively at a speed of 62 mm s^{-1} [BIE 00]*

In another publication [MIS 99], the authors studied the behavior of 34CrNiMo6 steel under friction against an aluminum sphere in a basic medium. They showed that anodic polarization (or passivation) of the material led to the generation of a surface oxide film that rapidly degrades under friction, whereas the same material showed improved resistance to wear in the cathodic phase. The wear mechanism reported by the authors consists of the growth of a weakly adhesive oxide film that deteriorates under friction, then eliminated as wear debris. The repetition of this mechanism of generation and degradation of the oxide film leads to the progressive consumption of the material and increasingly severe wear.

As a rule, the composition, stability and protective performance of steel passivation films depend on their composition, the type of electrolyte and the imposed potential.

Contrary to 34CrNiMo6 steel, with a passivation film which degrades easily under friction, stainless steels generate more stable surface passivation films which grant the materials better wear resistance [JIA 93, TAK 97c]. However, under particularly severe conditions (greater sliding speed and/or load), the passivation film can be destroyed and the material dissolved [PON 04].

2.8.2. Erosion-corrosion

In contrast to tribocorrosion which involves friction, erosion-corrosion is induced by the impact of a liquid either individually or mixed with an abrasive solid. Figure 2.46 shows the principle of a typical set-up designed for the study of erosion-corrosion. The polarized or non-polarized sample surface is subjected to the impact of projected corrosive liquid containing abrasive particles.

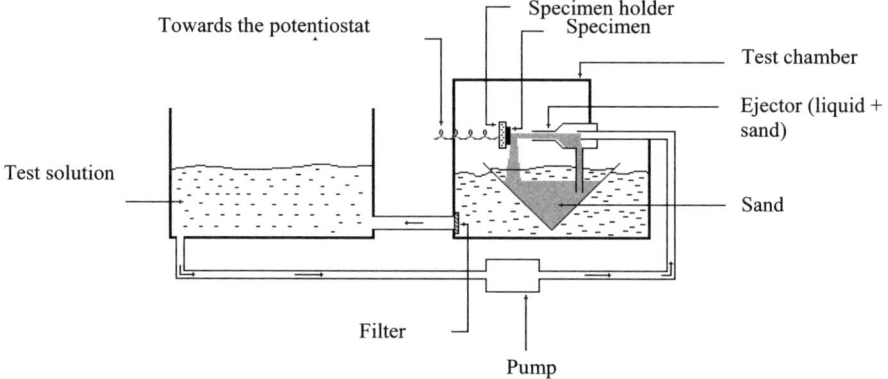

Figure 2.46. *Experimental set-up for the study of erosion-corrosion (adapted from [STA 06] and [WEN 95])*

As with tribocorrosion, many results in the literature show an important synergistic effect between corrosion and erosion, as well as the influence of the surface electrochemical potential on its resistance to wear [HAM 95, NEV 99, YUG 95]. This synergistic effect has recently been analyzed and modeled for many different materials. The results obtained have led to the establishment of numerous maps allowing the definition for each speed of projection: pH couple (or imposed electrochemical potential); the dominant mechanism underlying material removal (corrosion, erosion, etc.); as well as the extent of surface degradation [STA 04, STA 06].

In [NEV 99], the authors analyzed the behavior of gray cast iron under the impact of a projected liquid solution of NaCl at 3.5%, with and without the addition of silica sand particles.

Figure 2.47 shows the variation in the amount of material removed as a function of the projection speed of the corrosive liquid with or without the application of a cathodic potential to the sample (–0.8 V relative to a saturated calomel electrode).

The results clearly show that cathodic polarization enables the quantity of material removed to be reduced by around a factor of four.

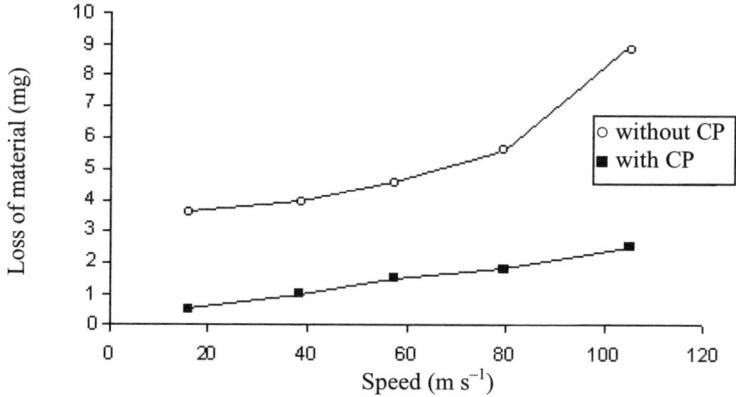

Figure 2.47. *Comparison of the total mass of material removed with and without cathodic polarization (CP) relative to the speed of projection of the corrosive liquid (NaCl solution at 3.5% and at 18°C) on the surface of a gray cast iron sample [NEV 99]*

Figure 2.48 shows the result obtained after the addition of silica sand to the NaCl solution. The quantity of material removed increases with the quantity of abrasive component used, with a plateau being reached when the concentration of silica sand is around 350 mg l^{-1}. The figure also shows the influence of the concentration of salt. If the solution is made less saline and the concentration reduced from 3.5 to 0.05%, the amount of material removed is halved. Applying a cathodic potential to the sample also provides very good results in this case.

This work also confirmed the existence of a strong interaction between wear and corrosion. A synergistic effect amounting to between 35 and 59% of the total material removed was found.

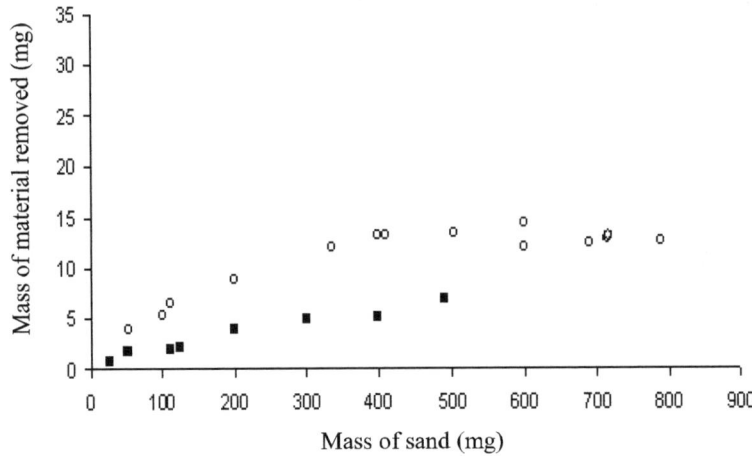

Figure 2.48. *The open circles (o) show the mass of removed material as a function of the amount of sand in the NaCl solution (at 3.5%) projected onto the surface of a gray cast iron (○). The solid squares (■) shows the same result when the solution has reduced salinity (NaCl at 0.05%) [NEV 99]*

It is important to note, however, that even if the cathodic polarization contributes in most cases to improved wear resistance, applying a high voltage for an extended length of time can lead to increased brittleness of the material due to hydrogen evolution. Moreover, when in contact with hydrogen, certain materials such as titanium form hydrides which are fragile and particularly brittle compounds. Such behavior was demonstrated in a study devoted to tribocorrosion of the TA6V alloy in a NaCl medium [BOUR 00]. Cathodic polarization was clearly shown to induce rapid degradation of the surface under friction.

Chapter 3

Materials for Tribology

3.1. Introduction

The choice of a pair of materials for a particular tribological application can only be made satisfactorily if the tribological system is completely known. This system includes the contact geometry and contact pressure, the type of motion, the relative sliding speed, the nature and thickness of any interfacial material (for example a lubricant) and the atmospheric environment in which the system is placed (temperature, humidity content and reactive chemical species); see Figure 3.1 for a schematic illustration.

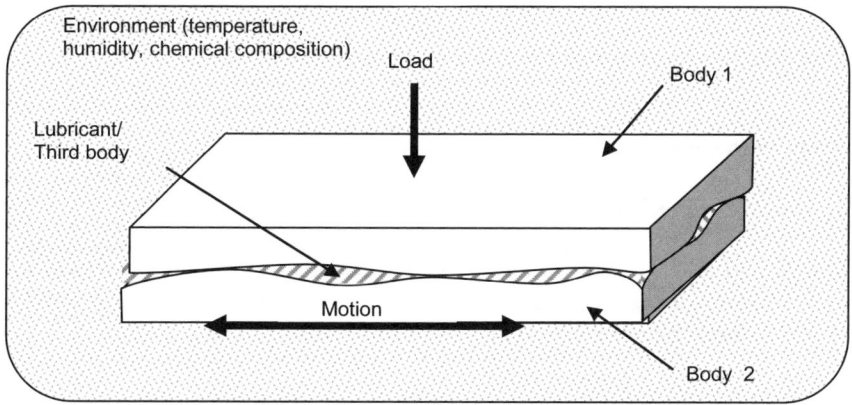

Figure 3.1. *The contact and its environment
(two solids in contact and in relative movement)*

The analysis of tribological phenomena requires, by nature, a multidisciplinary approach combining techniques derived from contact mechanics, surface physics and material and interface chemistry. As a result, there are no completely universal solutions to tribological problems. Each new case requires the very careful analysis of the specific problem under consideration in order to find the appropriate solution.

One of the basic rules of selection of material pairs is the so-called rule of "tribological compatibility" which states that materials which are mutually insoluble will exhibit a low tendency to adhesion and will constitute a good material friction pair. However, this rule should be used with caution as there are two important reasons why its validity is not always guaranteed:

– solubility is an intrinsic bulk property of materials, whereas friction depends primarily on surface characteristics such as hardness, crystalline structure, surface energy and topography;

– the solubility criterion does not take into account the chemical transformations (e.g. oxidation, phase transformation and generation of new compounds) or structural alterations (e.g. allotropic transformation, recrystallization and hardening) which can occur during friction and can modify the surface properties of the materials in contact.

When selecting a material for a particular tribological application, technicians and engineers can now choose from a wide range of materials including metals, polymers, ceramics, composites and lubricants, and can benefit from a variety of techniques for surface preparation, treatment and functionalization.

The present chapter will provide a review of the materials most commonly used in the field of tribology, covering materials used in their bulk state as well as those applied as coatings. The chapter also includes a description of the main techniques and processes for fabricating both thin and thick films and layers.

3.2. Bulk materials

3.2.1. *Metallic materials*

Metals and alloys make up the class of materials most commonly used in the field of tribology. Indeed, they are often found to be the most appropriate solution in many applications for cases involving dry or lubricated contacts.

3.2.1.1. *Iron-based alloys*

This class of materials comprises steels and cast irons. For these materials, a wide range of physical properties and mechanical and physico-chemical characteristics can be obtained through the careful variation of the material composition and the appropriate choice of thermo and thermo-mechanical treatments. Indeed, in most cases, ferrous alloys destined for tribological applications are subjected to preliminary surface treatments that are designed to improve their surface characteristics (as discussed later in section 3.3), and they are generally used with a lubricant.

3.2.1.1.1. Steels

The mechanical properties and resistance to wear of the different phases of steel differ very widely. For example, ferrite, austenite and pearlite are phases that are relatively soft (with a hardness ranging from 100 to 300 HV), and offer little resistance to wear. As a rule, the use of structurally homogenous materials such as ferrite and austenite is to be avoided, and it is preferable to use pearlite or a ferrito-pearlitic structure with carbide precipitation. Hard phases such as martensite and bainite (of hardness ranging from 800 to 1000 HV) are particularly well-suited when the surface is subjected to abrasive wear [LEV 94, MOOR 81].

The choice of steel microstructure used in a given application should take into account the type of stresses the material surface will be subjected to and the type of wear it is likely to undergo. In the case of abrasive wear, hardness should be the primary criterion [BERA 94] whereas in the case of impact, a microstructure that improves material ductility should be used. When there is a risk of adhesive wear, the surface composition should be tailored to that of the opposing material. More specifically, contact between opposing surfaces possessing the same microstructure should be avoided, particularly in the case of homogenous phases.

The mechanical characteristics of steel can be greatly improved by the adjunction of chromium, molybdenum, vanadium, manganese and nickel. Indeed, when added in small quantities (1 to 4%), these elements yield a finer microstructure and give the material greater hardness and wear resistance [MOOR 81].

Several varieties of steel intrinsically contain significant quantities of particular elements which lead to specific material characteristics. This is for example the case with manganese-rich steel (containing up to 15% manganese) which possesses high corrosion resistance and remarkable mechanical properties such as high toughness. This type of steel is therefore often used in the crushing mills found in quarries, cement works or mines. It is also used to manufacture rails, tools and even the bars used in prison cells. It could be argued that this material is particularly well-adapted to this latter application because the more one tries to saw it, the harder it becomes!

In the field of friction, strongly-alloyed varieties of steel are used for a great range of applications:

– tool steels or high speed steels (HSS) are used in applications requiring high resistance to wear and good mechanical properties withstanding high temperatures. These materials are used to manufacture wear rings, injection pistons and diverse types of machine tools. The mechanical and anti-wear characteristics of these tools are generally improved by the deposition of ceramic-type coatings such as titanium nitride, chromium nitride or titanium-aluminum nitride;

– stainless steels are used when good mechanical characteristics and adequate resistance to corrosion and wear are sought. These materials are commonly used for the fabrication of kitchen utensils and cutlery, as parts in pumps and for plumbing applications.

3.2.1.1.2. Cast irons

Cast irons typically present good frictional properties and excellent resistance to fatigue wear. There are two varieties of cast iron: white cast iron in which carbon is present in the form of cementite and gray cast iron in which carbon is mainly found in the form of either lamellar graphite (LG cast iron) or spheroid graphite (SG cast iron). White cast iron is hard and therefore presents very high resistance to abrasive wear. However, it is also brittle and difficult to machine. Gray cast iron, on the other hand, is more ductile. As well as offering good resistance to wear, it possesses excellent machining properties. It has the ability to absorb impacts and vibrations and to withstand thermal and mechanical wear. It is widely used in engine blocks, crankshafts, gears and other mechanical parts that are subject to cyclical stresses. Although the presence of graphitic carbon contributes to a reduction in friction at the surface of gray cast iron, its use in tribological applications often requires good lubrication.

3.2.1.2. *Superalloys*

Superalloys are complex materials which can be iron-, nickel- or cobalt-based. They contain significant amounts of additional elements such as aluminum, chromium, titanium, vanadium, molybdenum and tungsten. The composition of a typical superalloy is e.g. $Co_{30}\ Cr_{20}\ Ni_{20}\ Fe_{15}\ Mo_{10}\ W_5$.

These materials are characterized by good resistance to corrosion and high mechanical properties, even at high temperatures. Their excellent resistance to corrosion and oxidation at high temperatures is due to the presence of aluminum and chromium which form particularly stable protective oxides. They are, however, among the most costly metallic alloys.

The most commonly used super-alloys are nickel-based materials which can withstand temperatures up to 1100°C. In particular, they are used in the manufacture of gas turbines and aeronautic turbojet engines where they are primarily used as compressor disks or turbine blades. The second most important element in the composition of nickel-based super-alloys is aluminum. Small quantities of chromium are also added in order to enhance corrosion resistance. The adjunction of other elements, such as titanium and tungsten, also contributes to improved hardness properties.

3.2.1.3. *Copper-based alloys*

Copper and copper-based alloys can be used without lubrication when the contact pressure is not too high. Among copper-based alloys, bronze-lead and brass-lead are used for bushings, rails or as bearings materials. Copper-nickel and copper-aluminum alloys, while relatively ductile, exhibit better mechanical properties and better resistance to corrosion than their bronze or brass equivalents. As a result, they can be used under friction in corrosive media such as the marine environment.

Copper-aluminum alloys are less susceptible to corrosion than brass-based alloys when placed under strain in an ammonia environment, and are principally used as coinage alloys or for exchanger tubes in the potash and salt industries. After the addition of iron and nickel, foundry copper-aluminum alloys are used in the manufacture of pumps, turbines and heat exchanger plates.

The mechanical characteristics of copper-nickel alloys can be notably improved by the addition of other elements such as iron, aluminum or silicon. These alloys are commonly found in such applications as evaporators, heat exchangers, salt water pipes and hydraulic circuit coolers [CEN 92].

Finally, we note that bronze-based sintered materials can yield outstanding friction properties and excellent resistance to seizure when impregnated with oil or solid lubricants (such as graphite or molybdenum bisulfate). This last class of materials is widely used in friction rings (bushings, bearings, etc.).

3.2.2. *Polymers [CARREG 05, TRO 82]*

The surface energy of polymers is much lower than that of metals (from a few tens to a few hundreds of mJ m^{-2}, as opposed to the 1–3 J m^{-2} for metallic materials). This leads to low friction coefficients and good behavior in the case of dry friction. Conversely, these materials have low mechanical properties which limit their use to systems involving moderate contact pressure. The addition of reinforcement agents (such as fibers or ceramic particles) can significantly improve the properties of these materials and make them more suitable for mechanical applications.

Table 3.1 presents the main polymers used in tribology. Table 3.2 lists their characteristics.

Polymer	Formula
Polyethylene (PE)	$\left[CH_2 - CH_2 \right]_n$
Polytetrafluoroethylene (PTFE)	$\left[CF_2 - CF_2 \right]_n$
Polyoxymethylene (POM)	$\left[CH_2 - O \right]_n$
Polyamide (PA 6)	$\left[-NH - (CH_2)_5 - CO \right]_n$
Polyamide (PA 11)	$\left[-NH - (CH_2)_{10} - CO \right]_n$
Polyimide (PI-1)	
Polyetherethercetone (PEEK)	

Table 3.1. *Main polymers used in tribology*

	PEhd	Polyamide 6	Polyamide 11	POM	PTFE	PEEK	Polyimide
Melting point (°C)	130	220	185	175	330	335	388
Transitional vitrous temperature Tg (°C)	–110	55	30	–50	125	145	250
Maximum usability temperature (°C)	80	100	80	110	250	250	260
Young's modulus (MPa)	500–1200	2000	1000	3700	500	3700	2500
Resistance to traction (MPa)	18–35	70–85	56	62–70	25-36	100	75–100

Table 3.2. *Some characteristics of polymers used in tribology*

3.2.2.1. *High-density polyethylene*

It is usual to distinguish low-density polyethylene (LDPE), which is strongly ramified with little crystalline structure, and high-density polyethylene (HDPE), which is somewhat more crystalline, weakly ramified and has a larger elastic modulus. The terminology ultra-high molecular weight polyethylene (UHMWPE) is used when the average molecular weight is of the order 1 to 10 million. This type of polyethylene is widely used in tribology, particularly in applications such as:

– ski soles: UHMWPE combines high mechanical and anti-wear characteristics with remarkable hydrophobic properties and a low friction coefficient against snow (ranging from 0.10 and 0.01); and

– rubbing surfaces in joint prostheses for hips and knees (see Figure 3.24).

3.2.2.2. *Fluorinated polymers*

The most common fluorinated polymer is polytetrafluoroethylene (PTFE) or Teflon (the commercial name used by Dupont de Nemours). It is characterized by remarkable chemical inertness, making it resistant to corrosive substances such as hydrofluoric acid, nitric acid and caustic soda. It is therefore widely used in the chemical industry. PTFE is a non-stick material which behaves outstandingly under friction. However, its resistance to abrasion is limited and it is very susceptible to creep, which limits its use in mechanics. As an anti-stick material, PTFE is widely used as a coating for pots and frying pans, in molding devices (from shoe moulds to cake tins) or even as a coating on metallic parts used under friction. It is also used as a lubricant additive for many polymers and sintered materials.

3.2.2.3. *Polyacetal (polyoxymethylene: POM) and polyamide*

These materials combine a wide range of characteristics making them particularly well-suited for use in mechanical parts.

The main advantages of these materials are:
– their mechanical properties (high toughness, hardness and stiffness);
– their resistance to creep (for POM);
– their good frictional properties (low friction coefficient and excellent resistance to abrasive wear);
– their chemical stability in the presence of oils or greases;
– their good resistance to solvents (for POM); and
– their good resistance to fatigue.

In contrast to fluorinated polymers, however, POM and polyamide can be corroded by acids.

POM and polyamide are among the most commonly used polymers in mechanical manufacturing and they provide an attractive alternative to the use of metals. POM is therefore used to manufacture derailleur gears on bicycles, fixation clamps, zips or ski bindings. Polyamide material is used in precision gearings, mallet heads, pump parts or as spikes or soles for sports shoes.

3.2.2.4. *Polyimide*

Polyimide is an expensive material characterized by its excellent mechanical properties (which it retains to 250°C). Polyimide is characterized by a very low friction coefficient against most materials (0.01 to 0.1), presents excellent resistance to wear and exhibits very low creep. When bound to reinforcement agents, it can be used in high temperature tribological applications and under relatively heavy loads. It is routinely used for brake linings or gearings subject to intensive use.

3.2.2.5. *Polyetheretherketone (PEEK)*

PEEK is a material that combines remarkable mechanical characteristics with excellent high temperature behavior. Its melting point is 335°C and it can be used at up to 250°C. It has good tribological properties and exhibits good resistance to hydrolysis and to chemical products. PEEK can withstand most solvents, acids and bases and, as a result, is often used as a cladding material for electric cables placed in aggressive environments such as on oil rigs, in nuclear plants or in space. It is also used as anti-wear coating or for the manufacturing of high-quality components. On the downside, PEEK is much more costly than most common polymers.

3.2.2.6. *Friction and wear of polymers*

The dominant wear mechanisms are adhesive wear, abrasive wear and fatigue wear. In the case of metal/polymer pairs, several different types of behavior can be observed depending on the surface state of the metal. When the metallic surface is smooth, polymer wear mainly results from adhesion (soft polymers) or surface fatigue (hard polymers) [OMA 86]. Initially, fatigue wear induces the elastic deformation of asperities, and this is then followed by the appearance of cracks on the surface. As the network of cracks becomes larger, wear particles are generated which separate from the surface and are subsequently eliminated from the tribological circuit. Adhesive wear is characterized by a significant transfer of material from the polymer to the metal, which is in turn followed by a stationary state when the friction coefficient remains stable and wear is limited.

When the opposing surface is very rough, polymer wear occurs as a result of abrasion. As a rule, polymer wear is a function of the surface state of the opposing metal, as shown in Figure 3.2. Smooth surfaces undergo significant wear due to adhesion between the contacting surfaces. As roughness increases, the true contact area decreases and, consequently, so does the friction coefficient and the wear. However, when the roughness is too high, wear increases again due to abrasion against the metal asperities. Between these two extremes of surfaces that are smooth or very rough, an optimum value of roughness has been determined that minimizes the wear. In the case of friction of polyethylene against steel, two optimum roughness values corresponding to minimum wear have been reported: $Ra = 0.1$ μm [DOW 76] and $Rq = 10$ μm [BUC 81].

In contrast to metals, the resistance of polymers to abrasive wear does not depend on their hardness. Indeed, when placed in frictional contact with an abrasive surface, their rate of wear has been shown to be inversely proportional to the product AR_m, where A is the elongation at break and R_m the ultimate tensile strength [BRI 81, EVA 79].

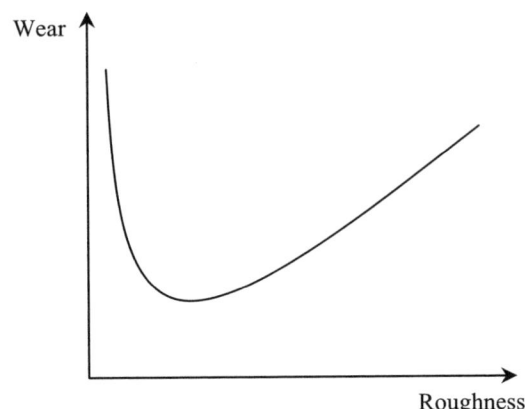

Figure 3.2. *Variations of polyethylene wear as a function of the roughness of the opposing surface. Wear minimum corresponds to a roughness of approximately Ra = 0.1 µm [DOW 76] or Rq = 10 µm [BUC 81]*

3.2.2.7. *Surface treatment of polymers*

There exist many types of surface treatments that can modify the physico-chemical properties of polymer surfaces. For example, flame or corona treatments activate the surface of a polymer by increasing the surface energy and so make the surface more receptive to the application of glue or paint. The mechanical and tribological properties of polymer surfaces can also be significantly improved through ion implantation. Figure 3.3 shows results of nano-indentation tests carried out on helium-implanted polypropylene/ethylene-propylene (PP/EPR) where significant hardening of the surface is observed. Measurements carried out at an indentation depth of 50 nm (for loading/unloading cycles recorded at a maximum depth of indentation of 50 nm) following an implantation of 10^{16} He^+ cm^{-2} yielded Young's modulus and hardness of 26.7 GPa and 6,720 MPa, respectively, to be compared with values of 1.9 GPa and 125 MPa for the non-implanted material.

When indentation tests were carried out at greater penetration depths, the ion implantation effect decreased but remained significant, as seen in Figure 3.3. This improvement of the mechanical characteristics of the surface was accompanied by a significant decrease in friction against 100Cr6 steel. A friction coefficient of 0.16 was obtained for the 10^{16} He^+ cm^{-2} dose implanted sample as opposed to 0.30 for the non-implanted sample [BRU 03].

The benefits of ion implantation on the surface mechanical characteristics of polymers have also been noted in the case of polyamide implanted with various elements such as Ne, C, N, O and Si. Figure 3.4 clearly shows that following neon implantation, the hardening of the polyamide surface increases proportionally to the amount of implanted ions [PIV 94].

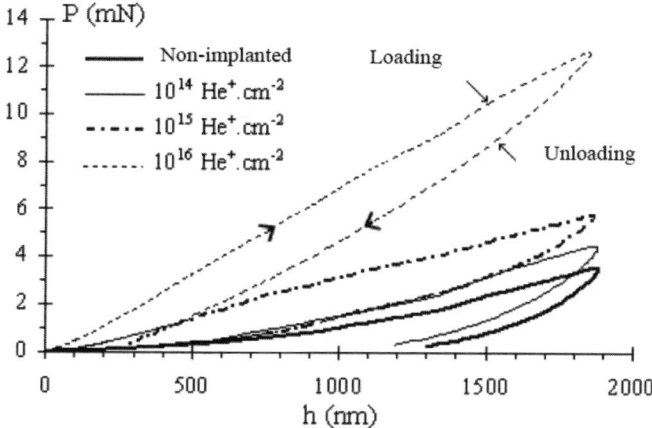

Figure 3.3. *Loading/unloading curve obtained by nano-indentation on PP/EPR implanted with varying helium doses [BRU 03]*

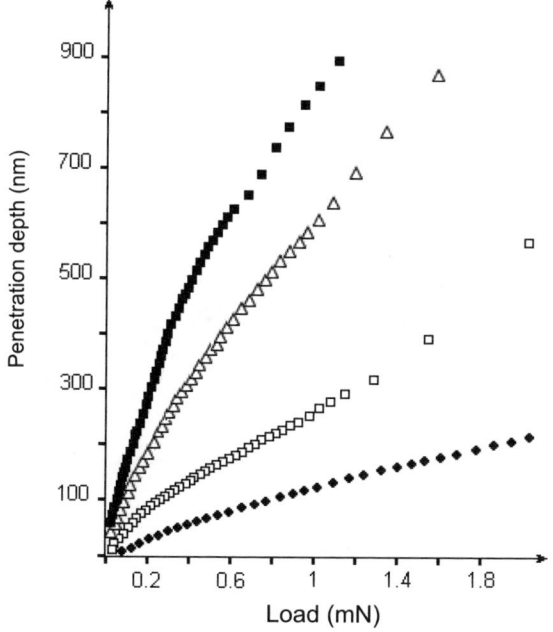

Figure 3.4. *Variation of the plastic deformation under indentation as a function of the applied load for neon-implanted polyamide (300 keV): untreated material or treated material with a dose of 10^{13} Ne cm^{-2} (■), 10^{14} Ne cm^{-2} (△), 5.10^{14} Ne cm^{-2} (□), 5.10^{15} Ne cm^{-2} (◆) [PIV 94]*

3.2.3. Composites

Polymers are rarely used for tribological applications on their own: they are usually combined with either reinforcing agents (such as glass, carbon or aramide fibers or ceramic particles) and/or lubricating agents (such as PTFE, MoS_2 or graphite) to form composite materials as shown in Figure 3.5. The goal of the composite is to improve the mechanical characteristics of the end material, as well as its friction coefficient and resistance to wear.

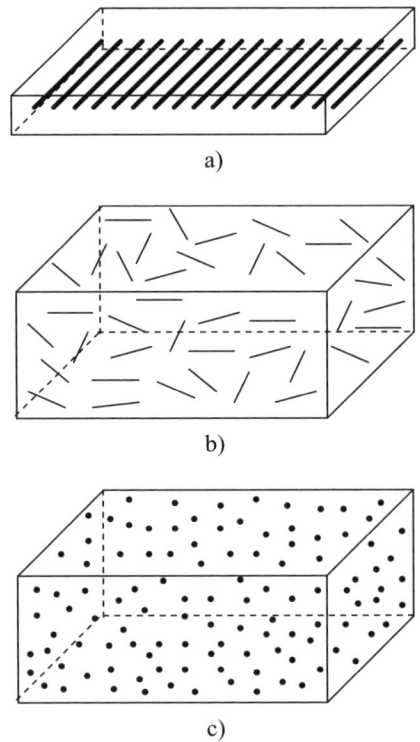

Figure 3.5. *Examples of composite materials in a polymer matrix. The reinforcing agents are: a) long fibers; b) short fibers; c) particles*

The addition of 30% glass fibers to a polyamide (PA6) can increase its resistance to traction from 80 to 150 MPa, its elasticity from 1,500 to 8,000 MPa and its resistance to compression from 60 to 150 MPa [ZYG 89]. The enhancement of these mechanical properties results in better resistance to wear, and this has been observed for a number of materials such as PEEK, polyamide, polycarbonate, PTFE and poly-acetal [FRI 95, FRI 98, LUZ 95, ZHANGS 97]. The wear resistance typically improves by a factor of 5–10.

In another study, volumes of PTFE from 5 to 85% were added to PEEK and the resulting composites were subjected to several friction tests. These showed that wear was minimized for a PTFE content of 15% [LUZ 95]. Other studies have determined an optimum fraction of added reinforcement in order to reduce wear [WANG 06]. A significant improvement to wear resistance at high temperatures (up to 260 °C) was also noted in the case of the PEEK/carbon fiber composite following addition of PTFE and graphite [PAU 96]. Other reinforcements such as silicon carbide can significantly improve PEEK's resistance to wear [ZHANGG 06]. In the case of polyetherimide (PEI), the friction coefficient against 100Cr6 steel was halved, while wear was reduced by between 60 and 80% with the addition of 5 to 20% carbon fibers [GUI 05].

It is important to note, however, that the tribological behavior of composites can be greatly affected by the presence of water and that it can be improved or degraded depending on the nature of the materials in question and of the stress they are subjected to. Water can corrode and increase the fragility of the base material-reinforcement interface but, on the other hand, it can also decrease the contact temperature and evacuate wear debris, which can be abrasive and contribute to material wear [HAN 87, YAM 04].

3.2.3.1. Friction materials

Friction materials are usually metal-matrix composites designed to transform kinetic energy into heat. The thermal conductivity of the elements of the metal-matrix (generally iron, copper or zinc) ensures heat dissipation, while the reinforcements (usually ceramics) give the material a high friction coefficient. These materials are used in brake pads and torque limiters.

3.2.4. Ceramics

The remarkable properties of ceramic materials make them particularly well-suited for tribological applications. These characteristics include:
– high values of hardness (thus high resistance to abrasion);
– a low coefficient of expansion (thus high dimensional stability);
– low reactivity (good resistance to chemicals); and
– an ability to withstand high temperatures.

Ceramics can be classified into two categories: oxides (Al_2O_3, MgO, ZrO_2, etc.) and non-oxides. Among the non-oxides, two broad classes of ceramics are particularly important in the fields of mechanics and tribology:

– nitrides (TiN, CrN, Si_3N_4, etc.); and
– carbides (TiC, SiC, ZrC, etc.).

Two other ceramic materials are particularly important as a result of their very high hardness: cubic Boron Nitride (cBN) and diamond.

Table 3.3 presents the characteristics of the main ceramics used in the field of tribology.

		Al_2O_3	ZrO_2/Y_2O_3	ZrO_2/MgO	SiC	Si_3N_4	SiAlON	BNC	Diamond
Melting point (°C)		2050	2590	–	2500	1900		3200	3825
Maximum use temperature (°C)		1800–1850	1500	–	1550	1400–1500	1500	–	–
Vickers hardness		1800–2000	1200	1200	2300	1450–1550	1500	4700	7000–10 000
Young's modulus (GPa)		300–400	200	200	420	290–315	290	660	1054
Poisson's coefficient		0.25–0.27	0.29	0.29	0.15	0.26–0.27	0.26	–	0.07–0.1
Fracture toughness (MPa m$^{1/2}$)		3–5	8–13	6–10	3.5–4	4–5	5.4	–	–
Linear dilation (× 10^{-6})		8–9	9–11	8–13	4–5	3.2	3	2.5–4.7	0.8
Thermal conductivity (W m^{-1} K^{-1})	20°C	29–32	1.9	1.9	180	18–20	20	60	2000
	500°C	12	2.1	2	68	18–20	20	–	–
	1000°C	9–10	2.2	2.2	40	18–20	20	–	–

Table 3.3. *Some characteristics of ceramics used in tribology*

Alumina-based ceramics are essentially used in the manufacturing of cutting tools, wear parts, sealing rings, grinding wheels or as electronic or heating appliance supports.

Zirconia ZrO_2 is characterized by its exceptional fracture toughness which is two to three times higher than that of alumina (see Table 3.3), making it a material very resistant to impacts. It can therefore be used in tribological applications requiring material ductility for which alumina would not be suitable.

At ambient temperature and under normal atmospheric pressure, zirconia crystallizes in a monoclinical structure which remains stable to 1,100°C, becomes tetragonal between 1,100 and 2,300°C and becomes cubic beyond 2,300°C.

The different phase changes of zirconia are reversible but they are accompanied by significant volumic variations. For example, because the tetragonal state is denser than the monoclinical state, the tetragonal to monoclinical transformation during cooling is accompanied by a volumic expansion of around 4%.

This phenomenon makes it problematic to design and manufacture engineering parts based on pure zirconia. Indeed, during cooling, such parts tend to crack and lose their mechanical resistance. In order to overcome this drawback, tetragonal (or cubic) zirconia is stabilized at low temperatures by doping it with 3 to 20% (molar percentage) of one of the following oxides: CeO_2, CaO, MgO or Y_2O_3. Once doped, this "partially stabilized zirconia" (PSZ) is then in a metastable state, and can recover its thermo-dynamically favorable (i.e. monoclinical) structure under the effect of temperature or mechanical stress.

Silicon and aluminum oxynitride (SiAlON) is another ceramic with remarkable properties: great hardness and fracture toughness, high thermal conductivity and very good resistance to wear. It is used in the fabrication of high-speed cutting tools, extruders and wire die plates.

Silicon carbide and silicon nitride offer excellent resistance to thermal shocks due to their low linear expansion coefficients and their high thermal conductivities. These characteristics make them particularly suitable for high-temperature mechanical and tribological applications. In an oxidizing environment, they become coated with a layer of partially hydrated SiO_2 which protects the surface from wear. Because of its superior mechanical properties when compared to steel (excellent resistance to corrosion and good behavior at high temperatures), silicon nitride is widely used to manufacture ball bearings for aeronautical applications, machine-tools and metrology. It is also found in other applications such as engine valves and in the manufacture of cutting tools.

Cubic boron nitride and diamond are characterized by their outstanding hardness and high fracture toughness. They are widely used in high-speed machining, particularly for hard and abrasive materials. Diamond is generally reserved for the machining of non-ferrous materials because of its oxidation at 700°C when in contact with iron and iron-based alloys. High-speed machining of steels is mostly carried out using cubic boron nitride because of its more stable chemical composition.

At the atomic level, ceramics are characterized by strong ionic-covalent bonds which yield good mechanical properties and excellent chemical stability.

Conversely, the strength of these inter-atomic bonds is also the origin of their fragility. In order to overcome this weakness, a second phase is often used. The materials thus obtained are called ceramic composites. An example of such materials is alumina reinforced with other ceramics such as titanium carbide or silicon carbide monocrystalline fibers called *whiskers*. These fibers, which are less than a micron in diameter and range between 5 and 20 microns in length, can significantly improve the mechanical properties of the material by preventing propagation of cracks (see Figure 3.6b).

Alumina can also be reinforced through the addition of tetragonal zirconia. When a crack appears within the material, the tetragonal zirconia grains in the vicinity of the crack are subjected to significant mechanical stress which triggers the tetragonal-monoclinical transition (see Figure 3.6c).

As noted above, this transformation is accompanied by an increase of the material volume of the order of 4%. As they expand, the zirconia grains impose additional compression stress on the crack, which retards its propagation (Figure 3.6c).

Another technique is based on the introduction of microcracks and voids in the ceramic during its manufacture, which leads to arrest crack propagation in a material [BOWE 86].

Materials for Tribology   125

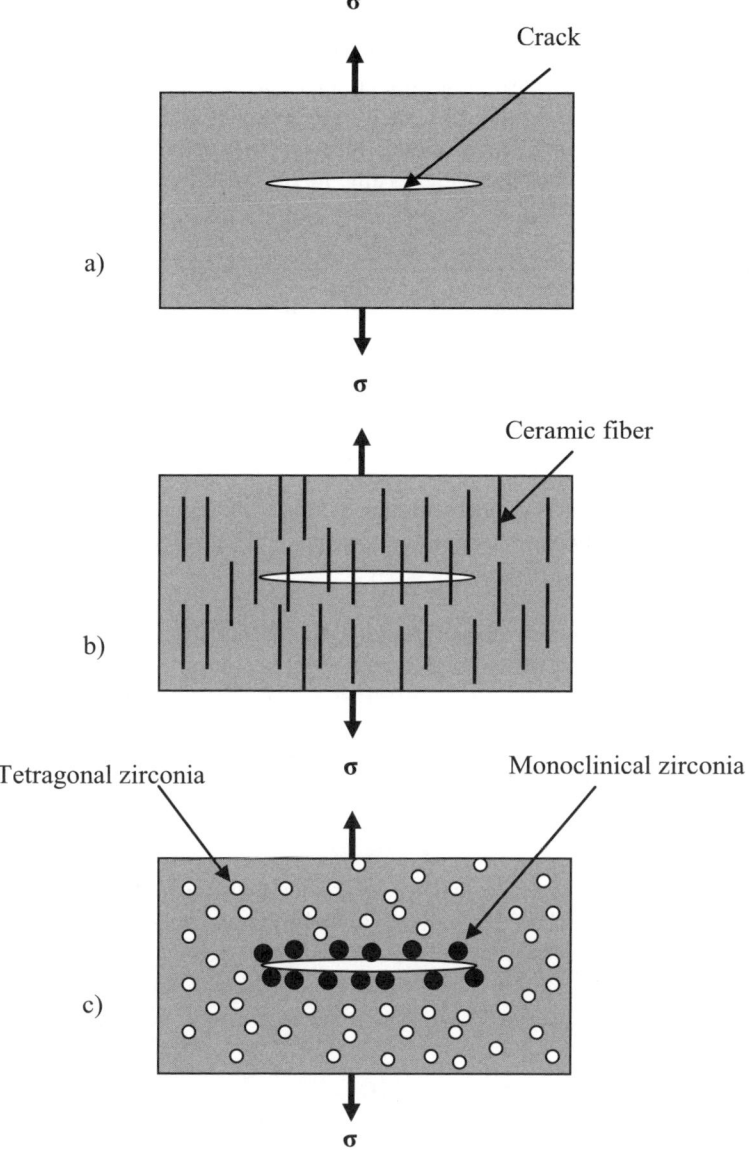

Figure 3.6. *Schematic representation of an alumina sample comprising a crack, subjected to stress σ: a) standard, non-reinforced alumina: the crack propagates as a result of stress; b) alumina reinforced with ceramic fibers preventing crack propagation; c) alumina reinforced with tetragonal zircone grains: the tetragonal to monoclinical transition of zirconia occurs as a result of the applied stress. This transformation is concurrent with a significant dilation of the material which prevents the propagation of the crack. Figure adapted from [BOWE 86]*

3.2.4.1. *Friction and wear of ceramics*

In order to understand the mechanisms of crack propagation and wear in ceramics, we consider the case of sliding contact between a sphere of radius R and a ceramic plane. If C is the length of a crack in the contact area and F the normal load applied to the sphere, it can be shown that crack propagation occurs when the normal applied load reaches the critical value F_c [ZUM 96]:

$$F_c = 1.1 \frac{\pi^{7/2}}{(1-2v)^3} \frac{K_C^3 R^{4/3}}{(1+C^*\mu)^3 C^{3/2} F^{2/3} E^{4/3}} \qquad [3.1]$$

where

$$C^* = \frac{3\pi}{8} \frac{4+v}{1-2v} \qquad [3.2]$$

and v and E are Poisson's coefficient and Young's modulus, respectively, μ is the friction coefficient between the sphere and plane and K_c is the fracture toughness of the ceramic.

By introducing two new terms (H and β, the hardness and the ratio of apparent and real contact area, respectively), equation [3.2] becomes:

$$F_c = \frac{2\pi}{1.12\beta(1-2v)} \frac{K_C RH}{(1+C^*\mu)C^{1/2} F^{1/2} E^{1/2}} \qquad [3.3]$$

When the applied load is equal to or greater than the critical value F_c, crack propagation occurs, which is generally followed by grain pull-out and rapid wear of the surface (see Figure 3.7).

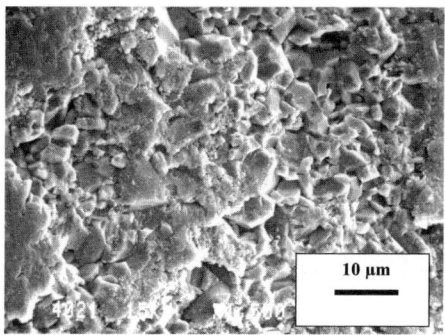

Figure 3.7. *Alumina surface showing significant wear of the material characterized by grain pull-out*

The mechanical surface characteristics of ceramics can also be significantly improved by treatments such as ion implantation or laser surface melting [POS 05, ZUM 00]. Laser surface melting involves depositing a layer of a given material onto the surface of a ceramic before irradiation by a high-energy laser beam, capable of melting it to form a new alloy or to precipitate new phases over a depth of a few hundred microns. Thus, starting from a suspension of powdered ceramics, HFO_2, TiN and ZrO_2 films have been deposited onto the surface of alumina samples with the addition of tungsten powder. After drying, these layers were heated to 1,500°C before being irradiated with a CO_2 laser beam. Friction tests carried out with an alumina sphere showed that the performed surface treatment resulted in a reduction in the wear by a factor of 4–8 and a reduction in the friction coefficient by a factor of 2 [ZUM 00].

As a general rule, the tribological behavior of ceramics is susceptible to humidity which can impact in two ways:

– the acceleration of crack propagation and, consequently, of material wear (in this case, the underlying physical mechanism is the breaking of inter-atomic bonds following their interaction with water molecules); and

– the tribochemical formation of a film (generally oxide/hydroxide) that is able to lubricate the contact and protect the surface.

Depending on the nature of the ceramic, the applied stress conditions and the level of humidity, these two phenomena can either occur simultaneously and compete, or occur independently. This potentially complex interaction accounts for the often contradictory results found in work published on the tribological behavior of oxide ceramics such as alumina and zirconia [CHENY 91, FIS 98, LAN 90, TAK 93b, TAK 93c].

For non-oxide ceramics, and more specifically silicon nitride and silicon carbide, all results point to a systematic reduction of the friction coefficient when the level of residual humidity increases.

The tribological behavior of these ceramics in the presence of humidity is governed by the following tribochemical reactions:

$$Si_3N_4 + 6H_2O \rightarrow 3\ SiO_2 + 4NH_3 \qquad [3.4]$$

$$SiC + O_2 + H_2O \rightarrow SiO_2 + CO + H_2 \qquad [3.5]$$

$$SiO_2 + 2H_2O \rightarrow Si(OH)_4 \qquad [3.6]$$

These reactions lead to the formation of a film consisting of a silicon oxide/hydroxide compound, which protects the surface and acts as a lubricant [LAN 90, TAK 94a, TAK 94b].

Figures 3.8 and 3.9 show the results of friction tests illustrating the beneficial effect of humidity on friction and on the wear of the Al_2O_3/Al_2O_3 and SiC/SiC pairs [TAK 94b]. The results of the tests presented in these figures characterize friction between a ceramic sphere of 5 mm in diameter and a plane sample at residual humidity levels of 20, 50 and 90%.

The figures show that in dry conditions or with low humidity, friction and wear are both significant. Moreover, when the applied load is increased, the level of wear occurring in the material becomes catastrophic [GAV 02b] in accordance with equation [3.3]. Indeed, this equation clearly shows that any increase in the normal applied load or friction coefficient brings about a reduction in the critical fracture load F_c, thus accelerating material wear.

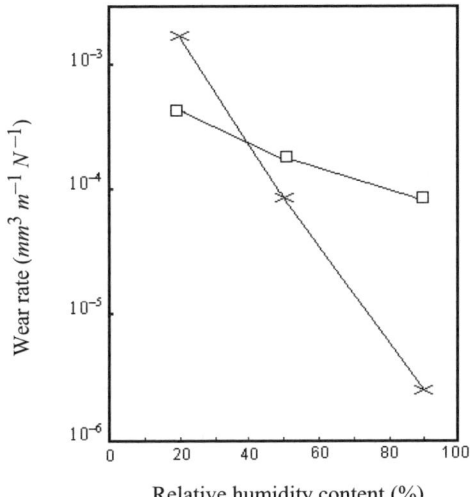

Figure 3.8. *Variations in the rate of wear of Al_2O_3/Al_2O_3 (□) and Si/SiC(×) couples/ pairs as a function of residual humidity. Tests were carried out with a reciprocating tribometer operating in the sphere/plane set-up and using a 5 mm diameter sphere under a normal applied load of 30 N. The sliding distance is 2 cm and the sliding speed is 2 mm s^{-1} [TAK 94b]*

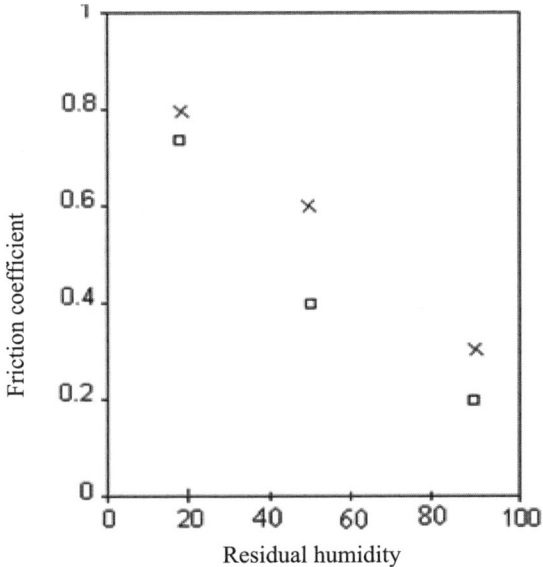

Figure 3.9. *Variations of the friction coefficient as a function of residual humidity: Al_2O_3/Al_2O_3 (×) and SiC/SiC (□) pairs (see Figure 3.8 for a description of the tests)*

Table 3.4 summarizes the data available in the literature in order to provide a qualitative overview of the influence of the applied load-humidity pair on the tribological behavior of ceramic materials when a protective and lubricating oxide/hydroxide film forms in humid conditions.

	Light load	Moderate load	Heavy load
Low humidity level **(< 30%)**	Moderate wear ($10^{-4} - 10^{-6}$)	Severe wear ($10^{-3} - 10^{-4}$)	Extensive wear ($> 10^{-3}$)
Moderate humidity level (30–70%)	Low wear ($10^{-6} - 10^{-8}$)	Transitional zone (low to severe wear)	Severe wear ($10^{-3} - 10^{-4}$)
High humidity level (> 70%)	Very low wear ($< 10^{-8}$)	Low wear ($10^{-6} - 10^{-8}$)	Moderate wear ($10^{-4} - 10^{-6}$)

Table 3.4. *Qualitative overview of the extent of ceramic wear as a function of the residual humidity and contact pressure for ceramic/ceramic pairs. Order-of-magnitude values for residual humidity (RH) as well as rate of wear (given in $mm^3 \ m^{-1} \ N^{-1}$) are given in brackets. The notions of light, moderate or heavy load are defined relative to F_c (equation [3.3]). The data presented only applies to ceramics which become coated with a protective, lubricating oxide/hydroxide film under humid conditions*

We also note that ceramics are sintered materials and therefore have varying degrees of porosity, impurities, agglomerates and vitreous phases at grain boundaries. These structural or chemical defects are precisely the sites from where cracks can develop before propagating between and within the ceramic grains. Materials manufactured under different sintering conditions and from ceramic powders of varying purity generally exhibit different tribological behavior. This partly explains certain results which can at times seem contradictory. The significant role played by these inter-granular vitreous phases needs to be stressed: indeed, at high temperatures (typically above 800°C), these secondary phases become viscous and radically alter the mechanical properties and tribological behavior of the ceramic material.

Concerning metal-ceramic couples, numerous studies have shown that the chemical reactivity of the metal is a determining factor in the tribological behavior of the pair under friction [BUC 81, BUC 94, PEP 76, TAK 92, TAK 93a].

Ultra-vacuum studies of friction between alumina and various metals (Ag, Cu, Ni and Fe, chosen as a function of their oxide stability which increases from Ag to Fe) were carried out and allowed us to establish that the adhesive contact between opposing surfaces could be perfectly correlated to the free energy of formation of the metallic oxides. This finding is in perfect agreement with the hypothesis that interfacial adhesion occurs *via* the generation of genuine chemical bonds between the metallic cation and the oxygen anion of alumina. When the metallic surfaces are exposed to oxygen, dioxides and trioxides such as $CuAlO_2$ or $NiAl_2O_3$ form at the interface during friction, resulting in an increase of the metal-ceramic adhesion and a subsequent increase of the friction coefficient [PEP 76].

In the case of transition metals, many authors have reported a strong correlation between the filling of the valence band (d-state) and the metal's reactivity. The surface is less reactive as the valence band is increasingly filled, leading to lower measured friction coefficients between the metals and various ceramics [BUC 94, MIY 82].

3.2.5. *Cermets*

A cermet is a composite composed of ceramic and metallic materials. Cermets possess both hardness and ductility as they combine a hard phase (the ceramics) and a soft phase (the metallic binding). When a crack forms within the ceramic, its propagation to adjacent grains is arrested by contact with the more ductile metallic phase (see Figure 3.10).

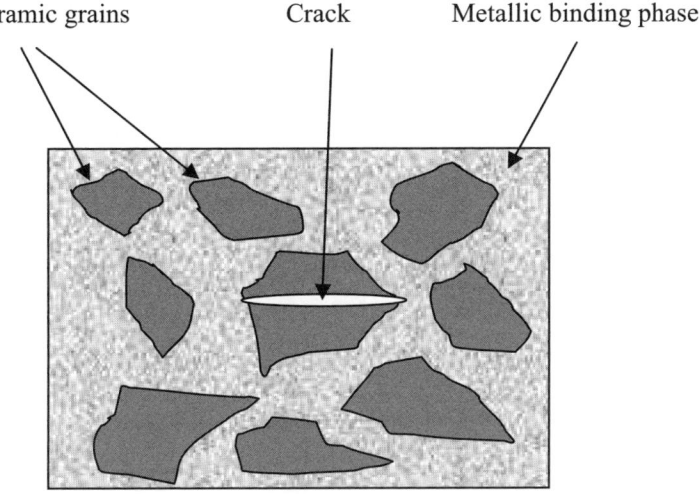

Figure 3.10. *Crack propagation through the ceramic grain is stopped by the metallic phase*

Cermets can be divided into two classes: tungsten-carbide (WC)-based cermets and other carbide-based cermets.

3.2.5.1. *Tungsten-carbide (WC)-based cermets*

These materials are produced from WC powders mixed into a cobalt binding agent, subsequently sintered at a temperature of about 1,500°C. At this temperature, the cobalt powder melts and the molten metal fills in the voids and thus binds the carbide grains together (see Figure 3.10).

The percentage of cobalt used is 5–20% by weight and the WC grains are generally 1–10 µm in diameter. The use of nanograins with diameters in the range of 0.1–1 µm makes it possible to generate materials presenting elevated hardness and improved wear resistance.

When the WC-Co carbide is subjected to friction, we initially see preferential wear of the binding phase (Co) followed by cracking and the subsequent loosening of the carbide grains [PIR 06, SHI 05].

The hardness and wear of the material strongly depend on the concentration of the binding phase. This is clearly shown in Figure 3.11 where we see an increase in wear and a decrease in hardness as a function of cobalt concentration. In contrast to the hardness, the fracture toughness increases with cobalt concentration from 8 to 14 MPa m$^{1/2}$ as the percentage of cobalt increases from 3 to 14% [PAST 87].

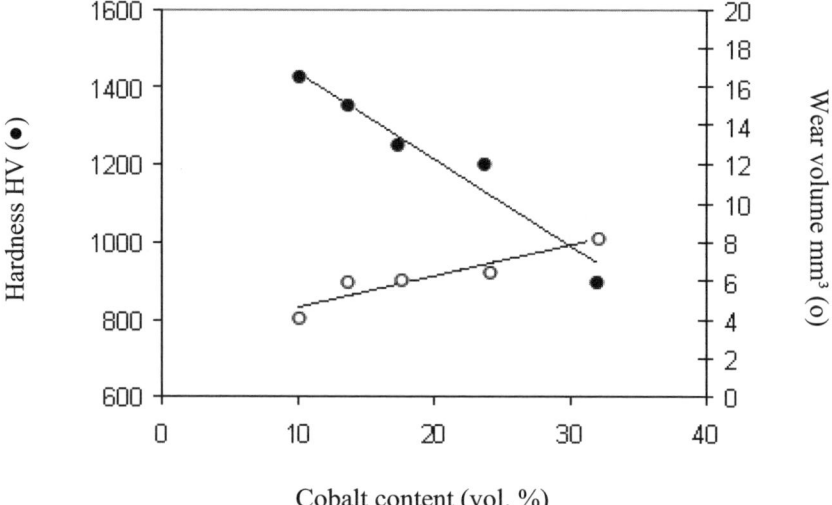

Figure 3.11. *Variation of the hardness and volume wear of the WC-Co cermet as a function of the binding phase concentration. The tribometer used is of the pin-on-cylinder type and the applied load is 40 N [PAST 87]*

Finally, we note that the wear resistance of tungsten carbide is improved through the addition of other carbides such as TiC, TaC and NbC or through the use of mixed carbides such as (W,Mo)C, (W,Ti)C or WC-(W,Ti)C-TaC.

3.2.5.2. *Other carbide-based cermets*

Examples of such composites may include TiCN-Ni, TiC-Ni, WC-TiN-TiC-Co, (Ti,Mo)CN-Ni, TiC-NiMo or TiCN-WC-Ni.

These materials are generally harder than WC-based cemented carbides and their resistance to wear is greater. When applied to cutting tools, they can significantly extend their lifespan and provide improved cutting quality [CEL 06, NIN 05, WAL 06].

As well as their application to cutting tools (for wood, metal or ceramics), cermets are also widely used as coatings on drilling heads, as well as hammering and shaping tools used in stamping, embossing and forging.

3.3. Surface treatments and coatings [BHU 91, CART 00]

There are many different techniques for surface treatments and coatings: they differ in their basic principles, their ease of practical implementation and the precise way in which they modify the surface of the treated materials.

Depending on the target application and its particular requirements, we can select techniques to modify the hardness, surface energy, friction coefficient, residual stresses, appearance or indeed any other mechanical, physico-chemical or aesthetic property of the surface.

Surface treatment and coating techniques can be classified into two categories:
– conversion techniques which modify the composition and/or structure of the surface to be treated; and
– deposition techniques which coat the surface to be treated with a thin layer of a given material.

3.3.1. *Conversion techniques*

3.3.1.1. *Anodic oxidation [BRA 91]*

Anodic oxidation or anodization is a surface treatment procedure mainly used for aluminum, titanium, niobium, magnesium, zinc and their alloys.

This treatment aims to develop a film on the surface of the material that is hard (to resist abrasive wear), refractive (to act as a heat screen) and chemically stable (to resist corrosion). Anodization is also often used as preliminary treatment before gluing or painting in order to improve the adhesion of the glue or paint to the substrate.

The material to be treated is immersed in the electrolyte which is usually a solution of sulfuric, chromic or phosphoric acid (although treatment in a basic solution is also possible) and it is connected to the positive terminal (anode) of a current source. Three types of reactions can occur at its surface (see Figure 3.12):

– dissolution: the metal is converted into metal ions in the solution;

– passivation: a thin film (generally a few nanometers thick) forms on the surface of the metal and arrests dissolution; and

– anodic oxidation (or anodization): the metal becomes coated with a porous film (generally an oxide/hydroxide combination) across which oxygen ions can diffuse and oxidize the substrate, increasing the thickness of the film (which can reach several tens of microns).

In order to predict which of the three reactions described above is likely to take place, Pourbaix diagrams (also known as potential/pH diagrams) are used as they allow the most thermodynamically favorable reaction to be identified (see Figure 3.13).

Materials for Tribology 135

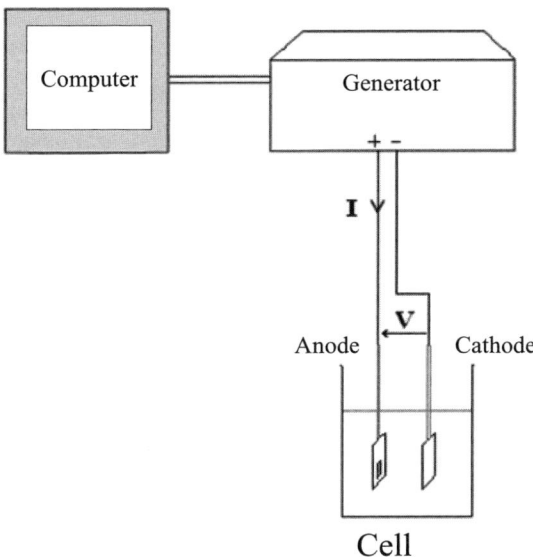

Figure 3.12. *Experimental set-up for anodic oxidation: the surface to be treated is the anode*

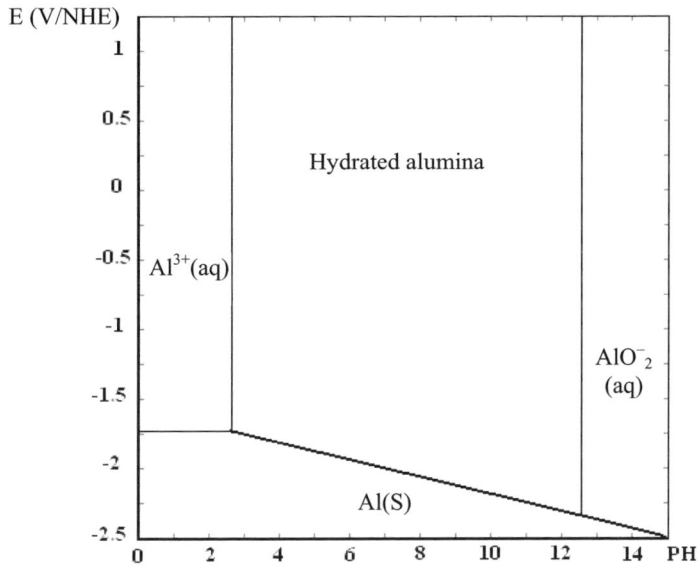

Figure 3.13. *Pourbaix diagram (potential/pH) for aluminum showing the domain of generation of alumina*

For the case of aluminum, the reaction mechanism can be divided into three stages.

– Stage 1: metal dissolution

$$Al \rightarrow Al^{3+} + 3e^- \qquad [3.17]$$

– Stage 2: dissociation of water and O^{2-} ion formation

$$2\,H_2O \leftrightarrows H_3O^+ + OH^- \qquad [3.18]$$

$$2\,OH^- \leftrightarrows H_2O + O^2 \qquad [3.19]$$

$$H_3O^+ \leftrightarrows H_2O + H^+ \qquad [3.20]$$

– Stage 3: formation of alumina Al_2O_3 (exothermic reaction)

$$2\,Al^{3+} + 3\,O^{2-} \rightarrow Al_2O_3 \qquad [3.21]$$

The structure of anodic alumina films consists of a hexagonal array of cells of varying depth, typically ranging from a few tenths to a few tens of microns (see Figure 3.14). These films are characterized by elevated hardness (600 HV compared to 100–200 HV for untreated aluminum alloys) and improved resistance to wear. However, as can be seen in Figure 3.15, anodized aluminum films are more brittle than non-anodized films.

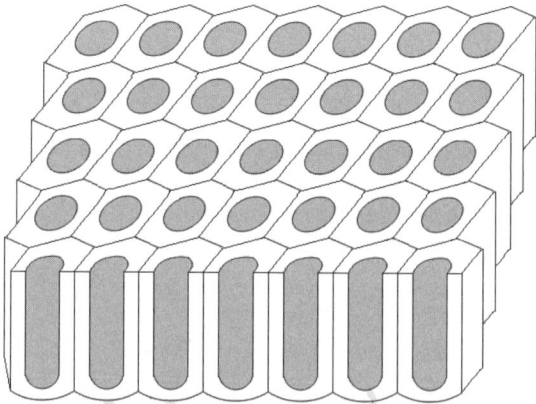

Figure 3.14. *Porous columnar alumina structure obtained through anodic oxidation (thickness: a few microns; cell size: a few nanometers)*

Materials for Tribology 137

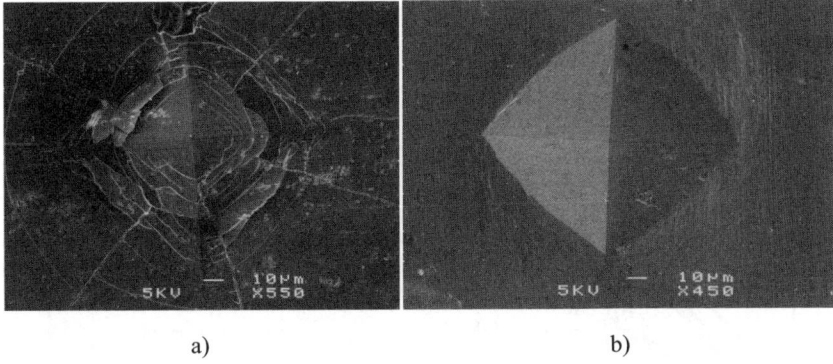

a)   b)

Figure 3.15. *Vickers indentations showing a) the brittleness of anodized aluminum relative to b) non-anodized aluminum which proves more ductile*

The morphological and mechanical characteristics of anodic films can be significantly improved by the use of pulsed signals (current or voltage) signals.

By imposing a current or voltage signal followed by a zero or weaker signal, overheating due to the exothermic nature of the anodization is significantly reduced. This prevents dissolution of the oxide film that is formed, reduces the time necessary for electrolysis and ensures that uniform film thicknesses are obtained.

Depending on the experimental conditions (polarization potential, temperature and composition of the electrolysis bath) alumina may naturally take on colors ranging from light gray to dark brown, with all possible intermediate nuances. In fact, it is possible to take advantage of the porous, columnar structure of alumina to add coloring agents in the form of organic or mineral pigments, allowing the manufacture of colored alumina (e.g. red, blue, green etc) for decorative purposes.

The columnar structure of alumina can equally be used as a lubricant reservoir, thereby increasing efficiency for applications in lubricated friction.

3.3.1.2. Ion implantation

Ion implantation introduces energetic ions into the surface layers of a solid.

An ion implanter is a device comprising a source that allows particular ions to be generated in the form of a beam (see Figure 3.16). An extraction system first removes the target ions and subjects them to an initial acceleration. The beam then passes through a separation magnet that filters out the ion species to select only the isotope with the desired mass/charge ratio. These ions pass through an acceleration column subjecting them to a voltage of several hundred kilovolts, and an electrostatic triplet then focuses the ion beam through a scanning system that consists of horizontal and vertical electrostatic deflection plates. This system allows the two-dimensional scanning of the sample and thus guarantees uniformity of the surface treatment.

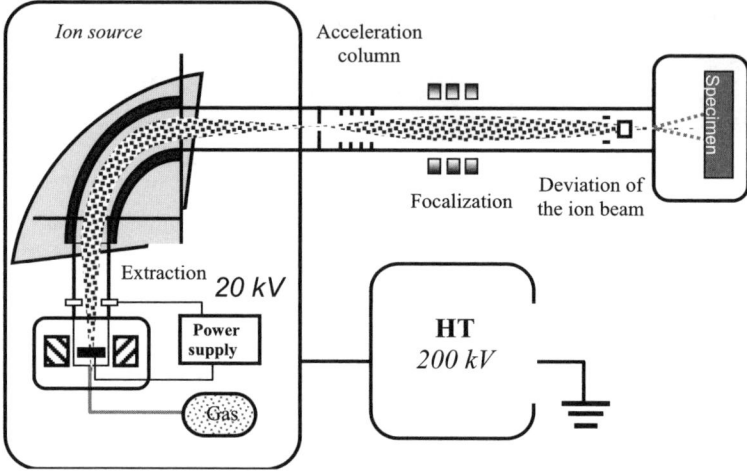

Figure 3.16. *Ion implanter*

The penetration depth of the ions can range from a few nanometers to a few microns, depending on their kinetic energy, their atomic number, their atomic mass and the nature of the material to be implanted.

During propagation in the solid target, an incident ion loses part of its energy through collisions with the target atoms. Atoms can then become displaced from their sites in the solid matrix and then displace other atoms; this can lead to a displacement cascade.

The mean energy loss for the implanted ion is expressed as the sum of two terms:

$$\frac{dE}{dx} = \left(\frac{dE}{dx}\right)_{nuclear} + \left(\frac{dE}{dx}\right)_{electronic} = -N\left(S_n(E) + S_e(E)\right) \qquad [3.22]$$

where:

– E is the energy of the implanted ion after propagation distance x within the solid;

– $S_n(E)$ is the nuclear stopping power (nuclear deceleration cross-section) arising from the interaction between the implanted ion and the atomic nuclei in the solid;

– $S_e(E)$ is the electronic stopping power (electronic deceleration cross-section) arising from the interaction between the implanted ion and the electrons of the solid; and

– N is the atomic density of the solid (expressed as atoms per cubic centimeter).

Knowledge of analytical expressions for $S_n(E)$ and $S_e(E)$ allows the calculation of the energy loss of the implanted ion throughout its path and allows the depth of implantation to be determined.

The total distance covered by the ion within the solid (see Figure 3.17) is given by:

$$R_L = \frac{1}{N} \int_0^E \frac{dE}{S_n(E) + S_e(E)} \qquad [3.23]$$

The distribution $C(x)$ of implanted ions perpendicular to the surface of the target is given by the following equation for a Gaussian distribution (see Figure 3.16):

$$C(x) = \frac{\phi}{\sqrt{2\pi} N \Delta R_p} \exp\left[-\frac{(x - R_p)^2}{2\Delta R_p^2}\right] \qquad [3.24]$$

where ϕ is the dose of implanted ions (expressed as atoms per cubic centimeters), R_p is the mean projected range of ions and ΔR_p is the standard deviation.

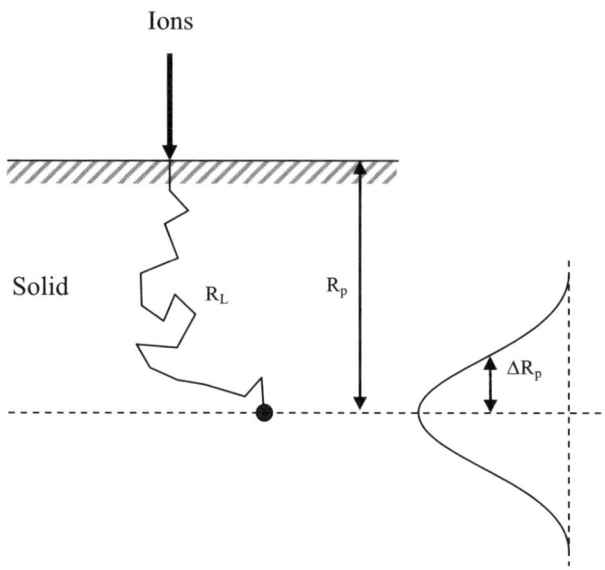

Figure 3.17. *Representation of the path taken by an ion implanted into a solid and (right) Gaussian distribution of the implanted ions*

Ion implantation induces two types of change to the surface of the solid: structural alterations (creation of disorder) and chemical alterations (modification of the chemical composition of the surface). These changes induce significant modifications in the physical, mechanical and physico-chemical properties of the implanted surfaces. In tribology, boron and phosphorus implanted into nickel have yielded significantly improved hardness (see Figures 3.18 and 3.19) and tribological behavior for the treated substrate [TAK 85, TAK 86, TAK 87]. Although the friction coefficient measured between pure nickel and a 100 Cr6 steel sphere of 5 mm in diameter was 0.95, it was reduced to 0.6 following ion implantation of 5×10^{16} B or P cm^{-2}, and to 0.3 following ion implantation of 2.5×10^{17} P cm^{-2} or 3×10^{17} B cm^{-2}. Nickel wear was thus reduced by over an order of magnitude following the implantation of heavy doses of metalloids.

Similar results have been obtained with many other metals: for example nitrogen ions implanted into steel [FAY 87], titanium [PIV 87] or the Ti6Al4V alloy [RIV 99]. We also recall that ion implantation has been successfully used to improve the mechanical and tribological properties of polymers (see section 3.2.2.7).

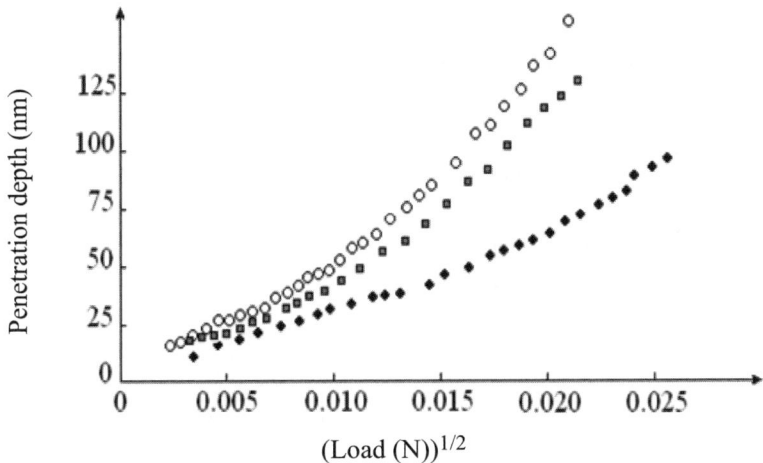

Figure 3.18. *Variation of the nano-indenter penetration as a function of the applied load for pure nickel and ion-implanted nickel with various doses of boron at 50 keV: (○) pure nickel; (■) Ni + 10^{16} B cm^{-2}; (◆) Ni + 2.3×10^{17} B cm^{-2} [TAK 87]*

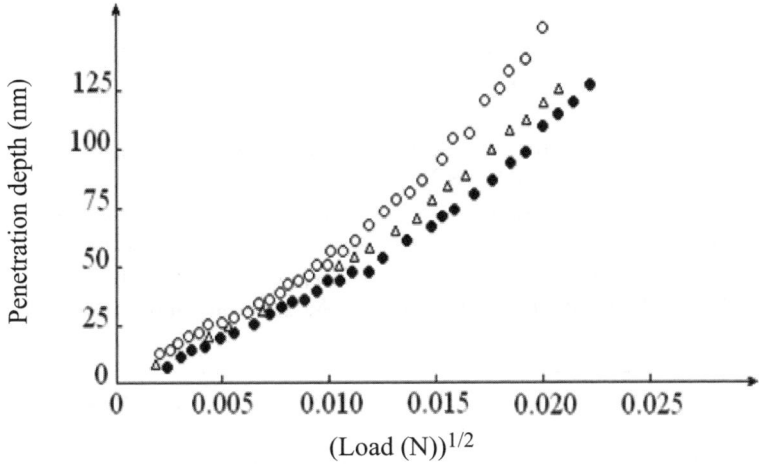

Figure 3.19. *Variation of the nano-indenter penetration as a function of applied load for pure nickel and ion-implanted nickel with various doses of phosphorus at 125 keV: (○) pure nickel; (△) Ni + 10^{17} B cm^{-2}; (●) Ni + 2.5×10^{17} P cm^{-2} [TAK 87]*

3.3.1.3. *Ion beam mixing*

Ion beam mixing is a technique where a film previously deposited onto a substrate is subjected to ionic bombardment. This treatment is generally carried out using argon, neon, krypton or xenon beams of energies 100–500 keV, and aims to induce substrate and film interdiffusion exchanges and the formation of a new alloy on the surface (see Figure 3.20).

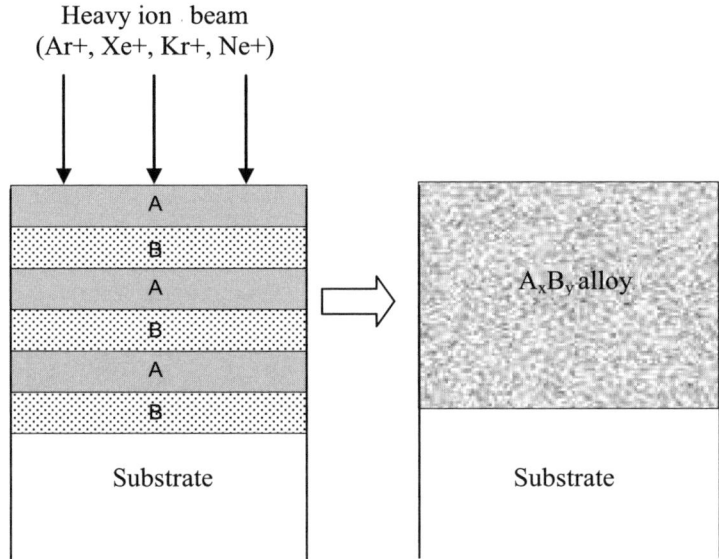

Figure 3.20. *Principle of ion beam mixing*

Ion beam mixing with multi-layered films of Fe/Ag [AMI 04], Ni/Mo [WEI 97], Si/C [RIV 95], Cr/V [BLAN 01] and Ti-TiN [HUB 01] has yielded the formation of new phases. Specifically, ionic bombarding of a nickel film deposited onto a silicon substrate under certain conditions has yielded a NiSi alloy with precipitation of the Ni_2Si phase [BOUS 05]. Alloys generated in this way generally possess remarkable mechanical [BLAN 01, WEI 97], tribological [WEI 97] and anti-corrosion [HUB 01] properties.

Metallic films have also been deposited onto ceramics and then subjected to ion bombardment, for example, the deposition of silver on Al_2O_3, ZrO_2 and Si_3N_4, or the deposition of niobium onto SiC. A significant improvement in the degree of resistance to wear was obtained in all cases [ERC 91].

3.3.1.4. *Thermochemical treatment*

Thermochemical treatments are mainly applied to steels [CONSTANT 92] and consist of enriching the surface of the material to be treated with certain metalloids. This process is referred to as carburizing when the metalloid in question is carbon and nitridation when nitrogen is used. Carbonitridation indicates that both elements are used and boronizing is used when boron is the metalloid involved.

The gaseous compound containing the carbon, nitrogen or boron, is reduced in contact with the material undergoing treatment and the metalloid is thus deposited in its solid state. As the material to be treated is generally heated to a temperature ranging from 500 to 1,000°C, carbon, nitrogen or boron will diffuse to a depth ranging from a few microns to several millimeters.

Carbon monoxide and methane are the gases used for carburizing, with the production of carbon resulting from the three following reactions:

$$2CO \rightarrow C + CO_2 \qquad [3.25]$$

$$CH_4 \rightarrow C + 2H_2 \qquad [3.26]$$

$$CO + H_2 \rightarrow C + H_2O \qquad [3.27]$$

Nitridation is obtained by decomposition of ammonia gas NH_3 in contact with the surface of the material which is heated to 500°C. Ammonia decomposes into hydrogen and nitrogen as:

$$2NH_3 \rightarrow 2N + 3H_2 \qquad [3.28]$$

Carburizing and nitdridation layers are characterized by significant hardness (of 700–1000 Vickers) and good residual compression stress.

Steels that have undergone this treatment are characterized by good resistance to wear, abrasion and fatigue.

Iron borides are very hard compounds obtained by diffusion of boron. Their hardness can range from 1,500 to 2,000 HV, which makes them particularly resistant to abrasion. Steel boronizing is usually carried out in a solid state using boron carbide.

3.3.1.5. *Transformation hardening*

Transformation hardening is another surface treatment that is also mainly applied to steels. It consists of heating the surface of the material to be treated to 800–1,000°C before cooling it down rapidly. Heating can be carried out under a flame, by induction or using an electron or laser beam. The depth thus treated can range from a few tens of microns to several millimeters.

As this surface treatment enables hardening of the surface and embedding of very high compression stresses, it can grant the material good resistance to both superficial fatigue and abrasion.

3.3.1.6. *Mechanical treatment*

Mechanical treatments consist of modifying the surface characteristics of materials through mechanical action. This can be carried out through sand blasting, hammering or shot peening. The aim of these mechanical treatments is to induce high compressive residual stresses into the surface layers, which can grant the material improved resistance to fatigue and abrasive wear.

Of all these mechanical treatment techniques, shot peening is the most widely used [CAS 91, FLA 91, LIE 87]. It consists of bombarding the surface being treated with a small spherical shot at a speed of 50–100 m s^{-1} (see Figure 3.21).

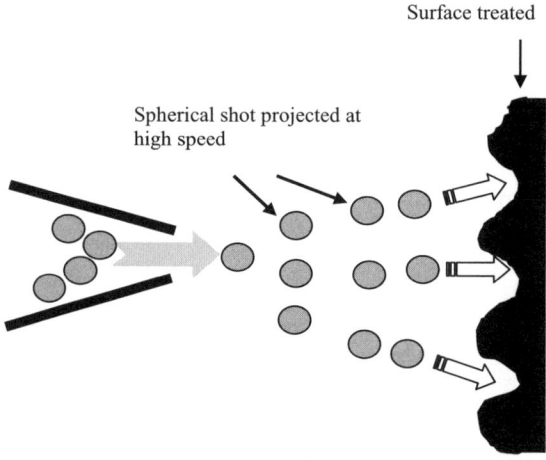

Figure 3.21. *Principle of shot peening*

The compressive residual stresses introduced into the surface are generally between –300 and –1000 MPa. The shots used are spherical glass, steel or ceramic particles with a diameter usually ranging from 0.02 to 2 mm. Steel shots are most commonly used in the industrial sector as the high material density enables the introduction of compressive stresses to significant depths, i.e. up to one or more millimeters.

Figure 3.22 shows the profile of the residual stresses introduced into the treated surface. It clearly shows the presence of a maximum of the compressive stresses at a depth of Z_0 below the surface.

In the case of a surface intended for friction, the mechanical treatment conditions will be chosen so that the depth Z_0 coincides with the zone to be subjected to the greatest shear stresses (see section 2.3.1, Figure 2.7).

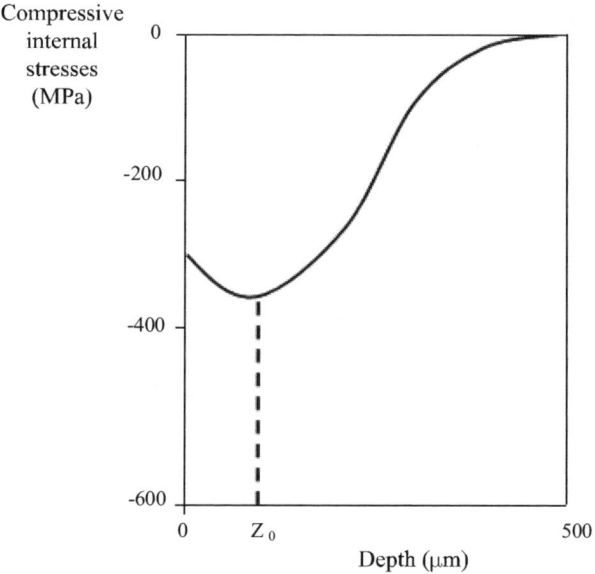

Figure 3.22. *Experimental result showing an example of the residual compressive stress profile introduced into steel by shot peening*

The kinetic energy (E) of the incident shot of mass m and density ρ is proportional to the square of the projection speed v and to the cube of the diameter d:

$$E = \frac{1}{2}mv^2 = \frac{\pi}{12}\rho d^3 v^2 \qquad [3.29]$$

By increasing the shot projection speed or the shot diameter, it is possible to increase the induced stress level as well as the depth affected by shot peening. However, the use of shot peening conditions that are too extreme can have adverse effects: if the shot projection speed, diameter or hardness is too high, or if the treatment time is too long or the degree of shot peening coverage is too high, this can lead to cracking of the material and rapid surface deterioration when exposed to mechanical stress.

Furthermore, however simple its implementation may be, shot peening can only be successfully performed provided some preliminary measures have been taken. Indeed, the operating conditions need to be tailored to the nature of the material being treated and to the mechanical stresses it is likely to be exposed to. It is therefore essential to be aware of the mechanical and metallurgical characteristics of the material in order to choose the type of shot and projection speed accordingly.

Note that it is also important for the treatment to be applied to the whole surface so that its mechanical properties are homogenous. The proportion of the surface of material to be treated is characterized by the degree of shot peening coverage N, which is defined as the ratio between the surface that has been impacted and the total area. Coverage of 100% corresponds to homogenous surface treatment.

As well as modifying the mechanical state of the surface layers of a material, shot peening also leads to high surface roughness. The degree of roughness increases with the diameter or projection speed of the shots, and/or with the softness of the material undergoing treatment. The final roughness of the surface is therefore that which is obtained after shot peening. In the case of tribological applications (under friction), a large (uncontrolled) roughness can cause significant tearing of material from the surface and even alter the efficiency of lubrication in the case of lubricated contact. A simple means of reducing shot peening-induced roughness is to perform a finishing shot peening process which, if carried out under moderate conditions, yields a surface having a smoother micro-geometric state.

3.3.2. *Deposition techniques*

Deposition techniques consist of the application onto a surface of a coating of a chosen material which is deposited either in the liquid or the gaseous phase. In order to ensure good adhesion of the coating onto the substrate, the surface of the material must be mechanically cleaned (through polishing, sand blasting, etc.), degreased and then activated using a physical or chemical process. Additionally, before depositing the coating onto the material, it is sometimes necessary to coat the surface with an undercoat which is designed to perform at least one of the following functions:

– facilitate the adhesion of the final coating onto the substrate;

– adapt the thermal expansion coefficients in order to reduce residual stresses within the coating (this is particularly true for high temperature processes); and

– act as a barrier layer when the aim is to avoid the diffusion of elements from the substrate into the coating and vice versa.

3.3.2.1. *Thermal projection techniques*

Thermal projection consists of melting the metal, ceramic or plastic-coating material and projecting it onto a surface in the form of droplets. These droplets are progressively superimposed and form deposits with thickness ranging from a few tens of microns to several millimeters. The principal thermal projection processes are based on the use of an electrical arc, a flame or a plasma torch (see Figure 3.23).

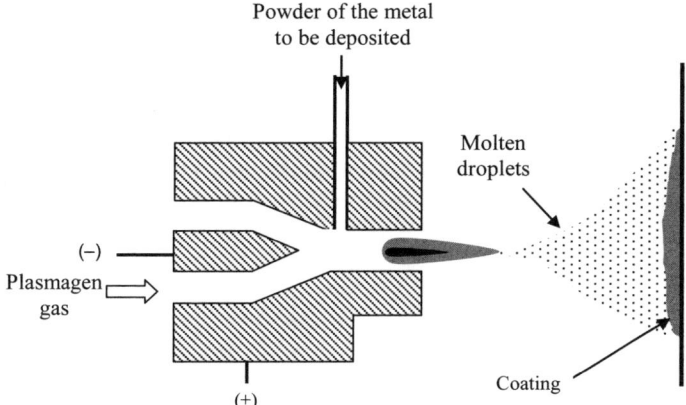

Figure 3.23. *Principle of thermal projection using a plasma torch*

The most widely used coatings in the tribological domain are alloys such as WC-Co, Al_2O_3-TiO_2, NiCrBSi, Cr_3C_2-NiCr and Cr_3C_2-NiAl. They are used in many applications such as the car manufacturing industry, where thermal projection is used on valves, piston rings, synchronization rings for gear boxes or clutch disks. In biomedical applications, certain prostheses such as artificial hip joints are covered with projected titanium or hydroxylapatite so as to facilitate their osteointegration and to encourage bone regeneration [COD 99] (see Figure 3.24).

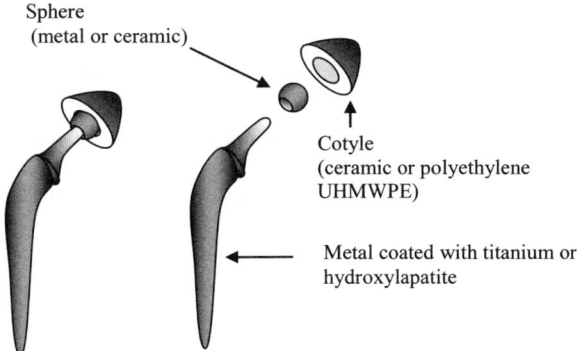

Figure 3.24. *Complete hip joint prosthesis*

3.3.2.2. Liquid-phase deposition techniques

There are two classes of liquid-phase deposition techniques:

– electrochemical deposition, which relies on an electric current or voltage source to induce the reaction that yields the formation of the coating; and

– chemical deposition, where the electrons required for the chemical reactions originate from the oxidation of a reducing agent contained in the bath, or from the oxidation of the substrate.

3.3.2.2.1. Electrochemical deposition

Electrochemical deposition, also referred to as electrodeposition or galvanoplasty, uses an electrolysis cell containing a bath in which electrodes are dipped (see Figure 3.25).

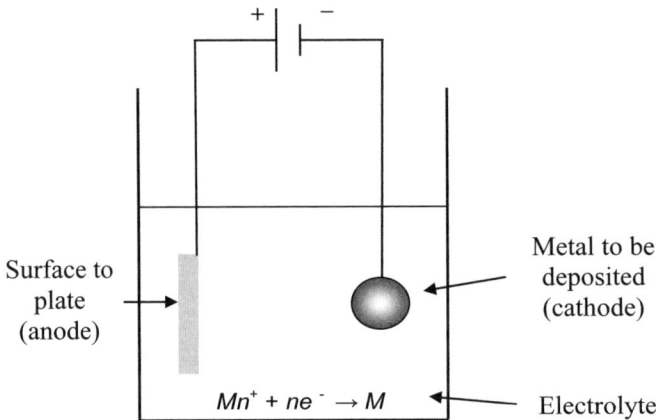

Figure 3.25. *Principle of electrodeposition*

The metal to be deposited is present in ionic form in the electrolytic bath. The electrodes are connected to a current source. The object to be coated is negatively polarized (cathode), while the anode consists of the metal to be deposited (except in certain cases, such as with gold, where an insoluble anode is used). The metallic ions in the electrolytic bath are reduced by contact with the cathode according to the following reaction:

$$M^{n+} + ne^- \rightarrow M \qquad [3.30]$$

At the same time, an atom from the anode passes into the solution:

$$M \rightarrow M^{n+} + ne^- \qquad [3.31]$$

These two inverse reactions compensate each other and allow the M^{n+} ion concentration to remain constant in the solution.

In an acidic medium (pH < 5), the reduction reaction occurring at the cathode is accompanied by the electrolysis of water and the dihydrogen evolution according to the reaction:

$$2H^+ + 2e^- \rightarrow H_2 \qquad [3.32]$$

Hydrogen diffuses within the coating and it often results in the brittleness of electrolytic deposits.

The electrical current densities used for electrochemical deposition usually range from 0.5–50 A dm^{-2}. These values can be significantly increased if pulsed currents are used. In this case, instead of subjecting the material to be treated to a continuous current, it is subjected to a pulsed current. The high current densities used yield increased deposition speed, higher density and harder coatings [MEN 00, NGU 98] and improved functional properties with regard to corrosion [CHAS 95, KON 97] and wear [NGU 98]. Moreover, pulsed current also yields coatings of homogenous thickness and avoids the development of excess thickness on the edges of the coated materials.

Figure 3.26 shows the form of the electrical signals commonly used. Two types of pulsed currents can be defined:

– simple pulsed currents consisting of cathodic pulses with a forward on-time Tc and a forward off-time Tr; and

– reverse pulsed currents when each cathodic pulse Tc is followed by an anodic pulse Ta.

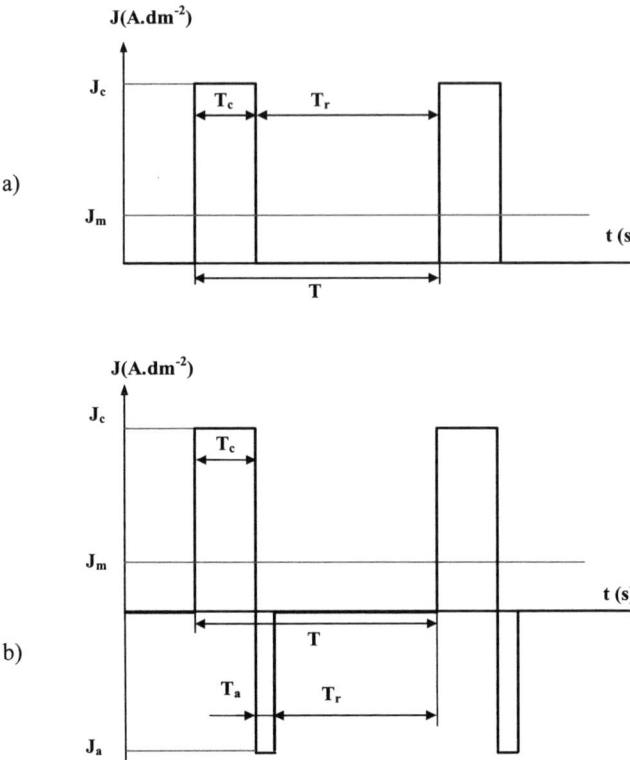

Figure 3.26. *Forms of pulsed currents used: a) simple pulsed current; b) reverse pulsed current (T: period (s); T_c: cathodic current application time (s); T_a: anodic current application time (s); J_c: peak cathodic current density (A dm^{-2}); J_a: peak anodic current density (A dm^{-2}); J_m: mean current density (A dm^{-2}); T_r: rest interval (s))*

The shape of the signal used can significantly impact on the morphology and characteristics of the deposited coating. Figure 3.27 and Table 3.5 show, for example, the case of chromium deposits for which three different types of morphologies and corresponding hardnesses were obtained under various deposition conditions [ADD 06].

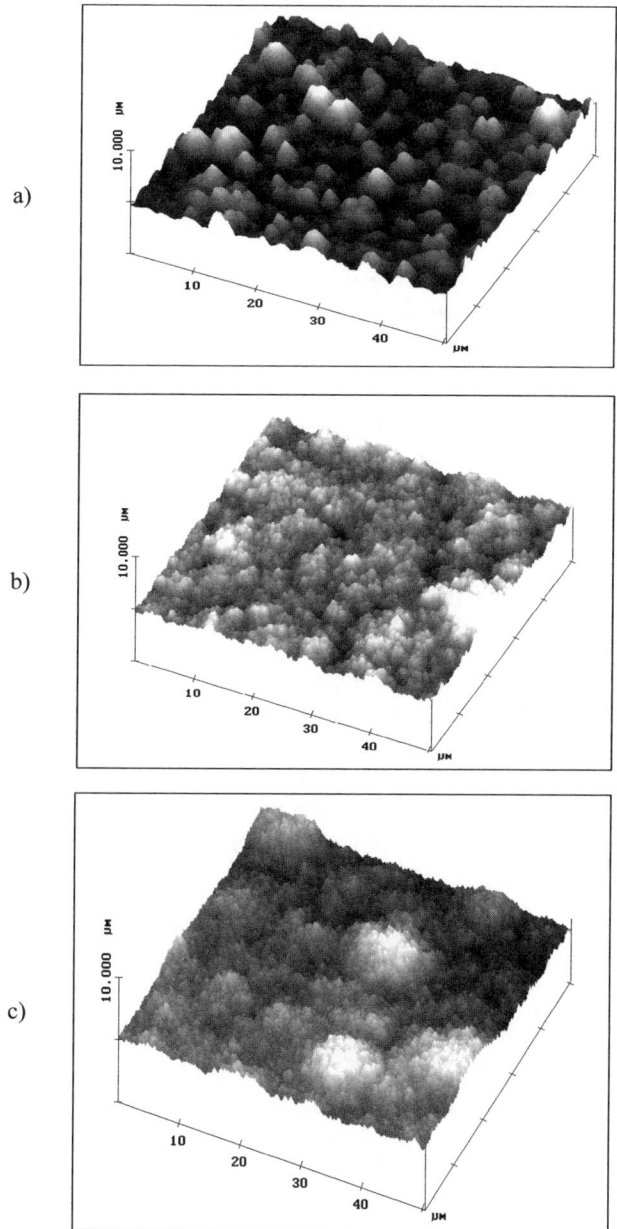

Figure 3.27. *Chromium coatings deposited onto steel (deposition conditions are listed in Table 3.5): a) large grains (hardness: 390 HV); b) medium grains (hardness: 557 HV); c) fine grains (hardness: 724 HV) [ADD 06]*

T (°C)	J_c (A dm^{-2})	T_c (ms)	J_a (A dm^{-2})	T_a (ms)	Morphology	Hardness (HV)	Figure
57.5	48	13000	48	40	Large grains	390	3.27a
57.5	36	13000	36	40	Medium grains	557	3.27b
57.5	36	5000	48	40	Fine grains	724	3.27c

Table 3.5. *Conditions of chromium deposition by reverse pulsed current with no forward off-time ($T_r = 0$ s): values for morphology and hardness [ADD 06]*

Electrochemical deposition also permits the generation of composite coatings consisting of a metallic base combined with hard particles (such as SiC, ZrO_2, Al_2O_3, diamond, etc.) or with solid lubricants (such as PTFE or MoS_2). These micro- or nano-size particles are kept suspended in the electrolyte using an appropriate agitator [BERC 03].

The electrochemical deposition of composites such as Au-PTFE [REZ 05] or Ni-PTFE [PEN 98] can significantly reduce the friction coefficient of a metal compared to its pure state. Moreover, the embedding of hard, usually ceramic, particles into the coating generally leads to improved resistance to abrasion and to increased material hardness. This is true of several compounds such as Ni-SiC [GAR 01, GROS 01], Ni-Al_2O_3 [GAN 04, SHR 01] or Ni-TiO_2 [LOS 99]. Figure 3.28 shows the variation of the hardness as a function of the volume fraction of ZrO_2 introduced in the case of the Ag-ZrO_2 composite. We see that the increase in hardness is proportional to the increase in the volume of embedded ceramic.

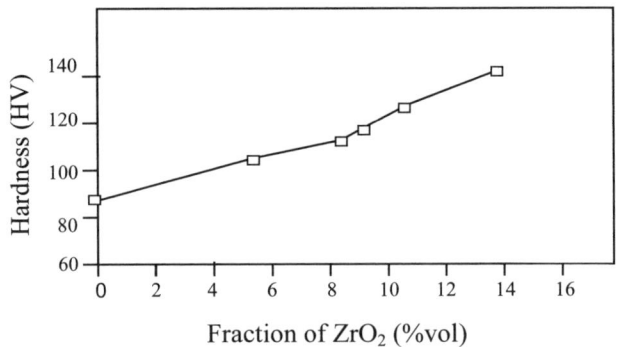

Figure 3.28. *Hardness of a Ag-ZrO_2 deposit as a function of the volume fraction of zirconia [GAY 01]*

In the case of the Ni-SiC composite, the mechanical characteristics of the material have been shown to depend on the volume fraction of SiC and its granularity (see Figure 3.29) [GROS 01]. Another study reported hardness values ranging from 650 to 1300 HV and an increased resistance to wear proportional to the percentage of siliceous carbide (see Figure 3.30) [ABD 06].

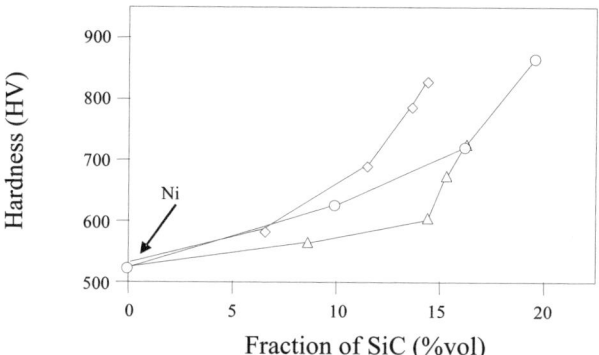

Figure 3.29. *Variation in the hardness of the Ni-SiC composite as a function of the size and volume fraction of ceramic particles. Mean diameter of grains: 1 μm (◇); 0.75 μm (○); 0.5 μm (△) [GROS 01]*

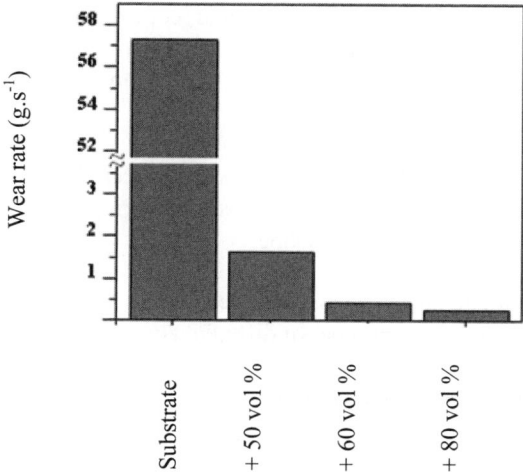

Figure 3.30. *Rate of wear as a function of the volumic percentage of silicon carbide in a Ni-SiC composite coating deposited onto an iron substrate. The wear tests were carried out using a pin/cylinder set-up. The cylinder was made of steel (of hardness 63 HRC) and its rotating speed was 30 rotations per minute (0.1 m s^{-1}). The applied load was 40 N and the fixed test duration was 0.5 hour [ABD 06]*

3.3.2.2.2. Chemical deposition

In the case of chemical deposition, the reduction of metallic ions does not require the use of a current generator. The necessary electrons are provided either by oxidation of a reducing agent present in the bath or by oxidation of the substrate which must necessarily be less noble than the metal to deposit. In the case of chemical nickel deposits, the hypophosphite ion is very often used as a reducing agent.

The two main reactions occurring in the solution are:

$$H_2PO_2^- + H_2O \rightarrow H_2PO_3^- + 2H^+ + 2\,e^- \qquad [3.33]$$

This reaction, corresponding to the reduction of hypophosphite ions, liberates electrons which will in turn enable the reduction of Ni^{2+} ions into metallic nickel:

$$Ni^{2+} + 2\,e^- \rightarrow Ni \qquad [3.34]$$

The total reaction is therefore:

$$Ni^{2+} + H_2PO_2^- + H_2O \rightarrow Ni + H_2PO_3^- + 2H^+ \qquad [3.35]$$

Commercially available baths can also be used to deposit nickel-boron or nickel-phosphorus alloys particularly onto steels and copper-based or aluminum-based alloys. These deposits are amorphous, characterized by good hardness (500–1000 HV) and good resistance to wear. They are usually deposited as thin films ranging from 15 to 40 microns.

The main advantage of the chemical deposition technique over the electrolytic technique is that it allows homogenous deposits covering the whole surface of the material to be treated, irrespective of shape.

As with electrochemical deposition techniques, chemical deposition can successfully be used to apply a number of different composite deposits.

3.3.2.3. *Vapor-phase deposition techniques*

A distinction is generally made between physical vapor deposition processes (PVDs) and chemical vapor deposition processes (CVDs) [GAL 02].

The characteristics of the coatings generated with these surface treatments depend on the technique chosen and the experimental conditions. Vapor phase deposition techniques have a wide range of applications: they may be applied as anti-reflection coatings for optical lenses, as thin films for electronic components or connectors or as decorative or anti-wear coatings.

3.3.2.3.1 Chemical processes (CVD and PACVD)

CVD deposits are the product of a chemical reaction between the heated substrate and one or several reactive gases that are introduced into the experimental reaction chamber (see Figure 3.31a).

As an illustrative example, we can see how titanium carbide and silicon oxide can be prepared according to the following reactions:

$$TiCl_4 (g) + CH_4 (g) \rightarrow TiC (s) + 4HCl (g) \quad\quad [3.36]$$

$$SiH_4 (g) + O_2 (g) \rightarrow SiO_2 (s) + 2 H_2 (g) \quad\quad [3.37]$$

The substrate temperature usually ranges between 500 and 1,500°C. These relatively high temperatures mean that this process cannot be applied to materials having a low melting point and, in the case of steels, to those having a tempering temperature higher than that at which the coating forms on the substrate.

Moreover, when the thermal expansion coefficient of the deposit and substrate are too different, the coating may exhibit high residual thermal stresses which can potentially cause it to peel off.

In order to overcome the drawbacks of classic CVD, other techniques have been developed. These techniques, known as plasma-assisted chemical vapor deposition (PACVD) and plasma-enhanced chemical vapor deposition (PECVD), are based on plasma generated by the creation of an electric discharge in the reaction chamber.

The high temperature required for the activation of the chemical reaction in the classic CVD technique is partly replaced by the action of the electrons accelerated within the plasma. The deposition temperature can therefore be significantly reduced to about 300°C (see Figure 3.31b).

The electrical discharge used is at a high frequency of 13.56 MHz in order to allow deposition of insulating films. Indeed, the deposition of such films would not be possible using a direct current as it would rapidly stop the discharge.

This technique can be applied to yield a number of metallic and ceramic materials with remarkable mechanical and tribological characteristics [PAW 03, RIC 94].

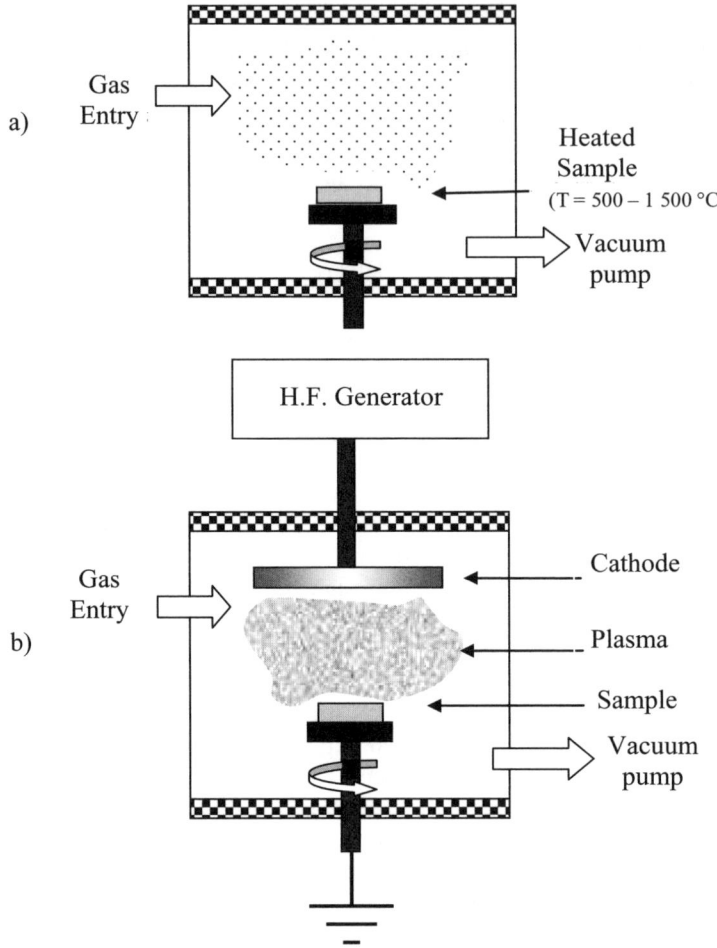

Figure 3.31. *Vapor phase chemical deposition:
a) CVD; b) PACVD*

3.3.2.3.2. Physical processes

PVD refers to several types of processes operating under low pressure conditions (ranging from 10^{-4} to 10^{-1} Pa) and at low temperatures (between 100 and 500°C).

A distinction is generally made between three types of physical processes:
– thermal evaporation;
– ionic deposition; and
– sputtering.

Thermal evaporation

This technique consists of vaporizing the source material to be subsequently deposited onto the substrate by heating it in a vacuum (usually ranging from 10^{-2} to 10^{-3} Pa). The atoms that are evaporated are then progressively deposited as a film onto the surface of the substrate to be treated. The most commonly used vaporizing sources are wires or foils heated with an electrical current. The material to be deposited can also be sublimed under the impact of an electron beam [GAL 02, RIC 94]. Moreover, it is possible to significantly improve the density of the film as well as its adhesion and its mechanical properties by carrying out ionic bombardment during its growth: an ion beam (generally argon) is then aimed at the substrate and bombards the film as it grows. This particular technique is known as ion beam-assisted deposition (IBAD) (see Figure 3.32).

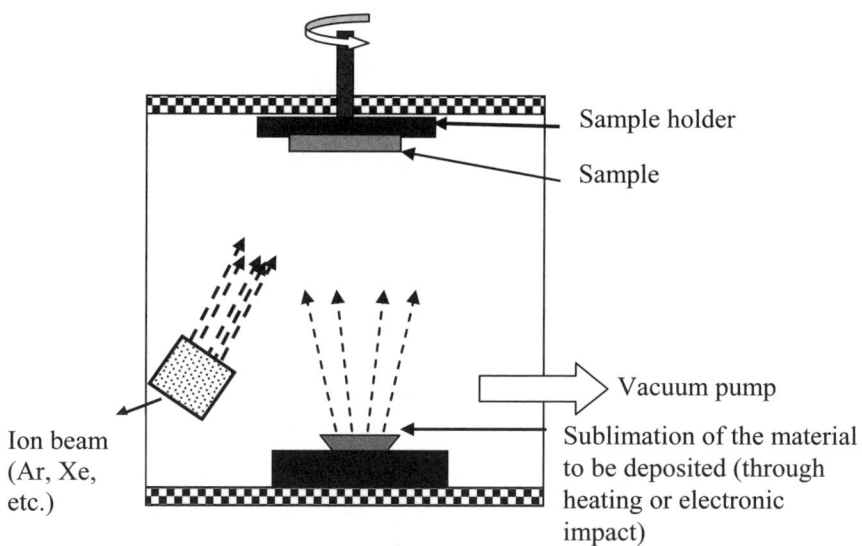

Figure 3.32. *Ion beam assisted deposition (IBAD)*

Prior to thermal evaporation the ion beam is used to clean the surface of the substrate through ionic etching, which consists of removing the contamination layer of surface oxides. This significantly increases the reactivity of the surface and so ensures better adhesion of the deposited films.

Ionic deposition

Ionic deposition (or *ion plating*) consists of evaporating the deposition material inside a chamber and subsequently creating a discharge by negatively polarizing the specimen to be treated. The ionized atoms are attracted to the surface of the specimen, which is subjected to ionic bombardment throughout the deposition process. In addition, reactive gases such as oxygen, nitrogen or hydrocarbides can also be injected into the reactor to form oxides, nitrides or carbides, respectively. This technique of ionic deposition is referred to more specifically as reactive ion plating (RIP).

At elevated deposition speeds, this technique allows the formation of dense coatings that exhibit good adhesion to the substrate.

Sputtering

This technique consists of applying a continuous voltage of a few hundred volts between the material to be deposited (cathode) and the material to be coated (anode) in the presence of argon in a vacuum ranging from 10^{-1} to 10^{-2} Pa. This results in the ionization of the accelerated argon atoms that bombard the cathode, causing the ejection of surface atoms which leave the target with high kinetic energy and are deposited onto the substrate (see Figure 3.33).

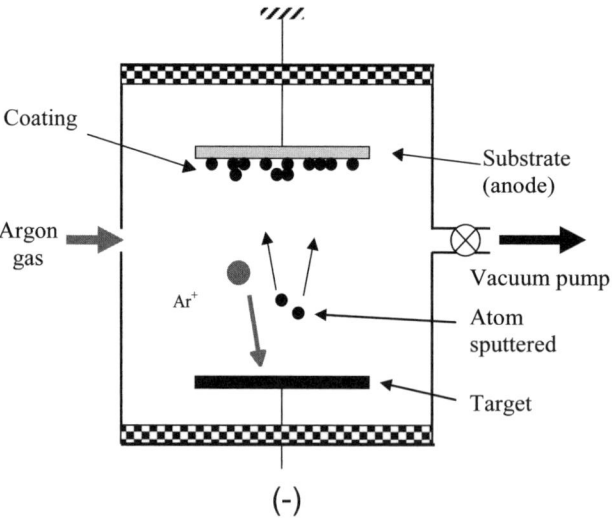

Figure 3.33. *Principle of sputtering*

Reactive gases other than argon can also be injected into the reactor. For example, oxygen or nitrogen can be used, and these gases combine with the atoms ejected from the target to form oxide or nitride films. This technique is referred to as reactive sputtering and allows films of TiN, CrN, AlN, SiO_2 or TiON to be easily obtained.

After deposition of the coating, it is possible to negatively polarize the sample being treated in order to attract argon ions and thus submit it to ionic bombardment. This treatment can also be performed during film preparation. It effectively increases the density of the deposition material and yields improvement in the interface quality and in the adhesion between different layers for multi-layered material deposition. This has been clearly demonstrated in the case of multilayered TiN/AlN films [THOB 99, THOB 00]. In addition, ionic deposition can also be assisted by a beam of argon ions that bombard the film during growth. Many alloys such as TiB_2, SiC or NiTiN have been produced in this way and deposited onto silicon, steel and TA6V substrates [RIV 00].

However, applying a continuous voltage to the cathode does not allow the sputtering on targets made of insulating material, because sputtering is rapidly arrested by the accumulation of surface charge. This problem is solved using a radio frequency system where an alternating voltage is applied to the cathode. With this technique, argon ions bombard the surface of the target material and eject the surface atoms during the negative-going polarization phase. When the polarization is reversed, the cathode attracts the electrons which then neutralize the positive surface charge.

The deposition speed depends on a number of parameters such as the sputtering yield of the target material, the kinetic energy of the argon ions and the reactor geometry. The deposition speed can be greatly increased by fitting the sample-holder with a magnetron (Figure 3.34) in order to apply a strong magnetic field using magnets placed behind the target. Electrons therefore follow the magnetic field lines and follow a helical path. This significantly increases the length of their trajectory and, as a result, the number of their collisions with argon atoms. The ensuing increase in the number of argon ions leads to greater sputtering and enhanced deposition speeds.

Films produced through cathodic sputtering are generally characterized by a columnar structure with a morphology that depends on the pressure and on the ratio between the deposition temperature (T) and melting point of the material deposit (Tm).

Figure 3.35 shows a schematic representation of typical microstructures obtained. For a relative temperature (T/Tm) between 0 and 0.4, the structure obtained is porous and the columns are very thin (see Figure 3.35a). When the relative temperature is between 0.6 and 0.8, the columns join together and the structure of the material is denser (see Figure 3.35b). At higher relative temperatures (T/Tm ranging from 0.8 to 1), we observe recrystallization and the growth of grains (see Figure 3.35c). For relative temperatures in the range 0.4 to 0.6, the structures are poorly-defined. It is important to note, however, that these temperature ranges can vary slightly as a function of the pressure [MES 84, THOR 74].

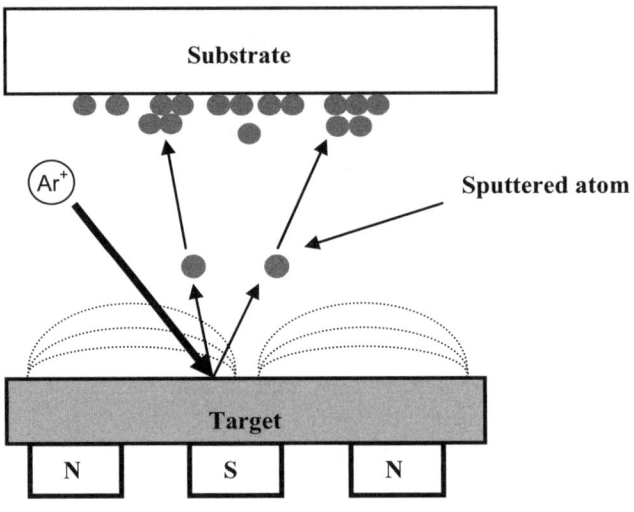

Figure 3.34. *Principle of the magnetron. Under the effect of the magnetic field, the electrons take on a helicoidal trajectory. The atoms of the target are sputtered under the impact of the argon ions and are deposited onto the substrate*

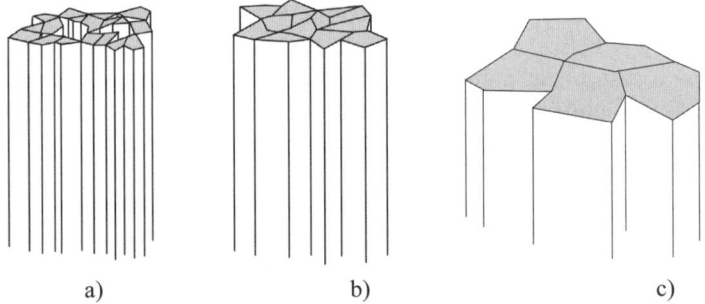

Figure 3.35. *Schematic representation of the microstructure of a film deposited by sputtering: a) porous microstructure made up of fine columns; b) wider columns of a less porous structure; c) recrystallization and growth of grains*

Sputtering set-ups always place the target material and sample to be coated opposite one another (see Figure 3.33), allowing for the generation of column deposits perpendicular to the surface of the sample. Modifying the angle of the sample-holder to a certain degree relative to the target makes it possible to achieve oblique films. However, use of this glancing angle deposition (or GLAD) technique is currently somewhat limited [BUZ 04, SATOM 00, SET 99].

Experimentally, it is observed that there is always a difference between the angle between the normal to the surface and the direction of incident particle flux (α) and the angle between the normal and the direction of column growth (β) (see Figure 3.36). The angle α is generally between β and 2 β.

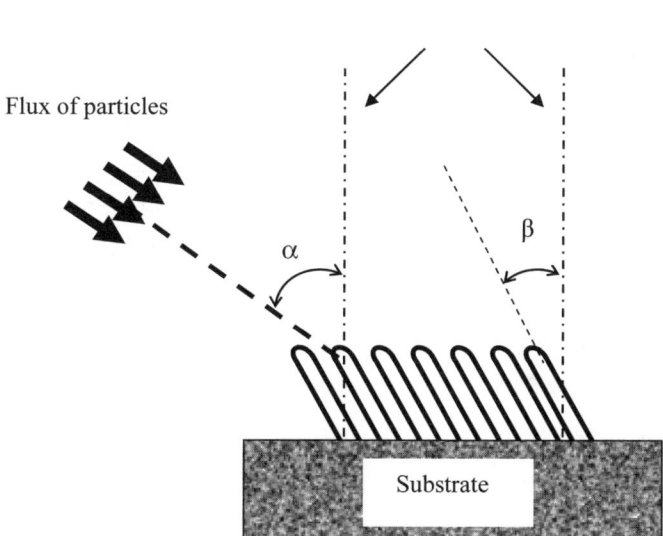

Figure 3.36. *Definition of the incident particle angle (α) and of the glancing angle (β) of the columns relative to the substrate (adapted from [LIN 03a])*

Figure 3.37 shows the microstructure and surface morphology of some oblique chromium films deposited onto silicon for various angles α. It is clearly seen that the surface topography can be modified by changing the incident angle of the primary particles. It is therefore possible to obtain surfaces of varying porosity and roughness and with a tailored distribution of hills and valleys.

The possibility to produce surfaces with topographically engineered surfaces makes this technique a very important tool for tribological applications.

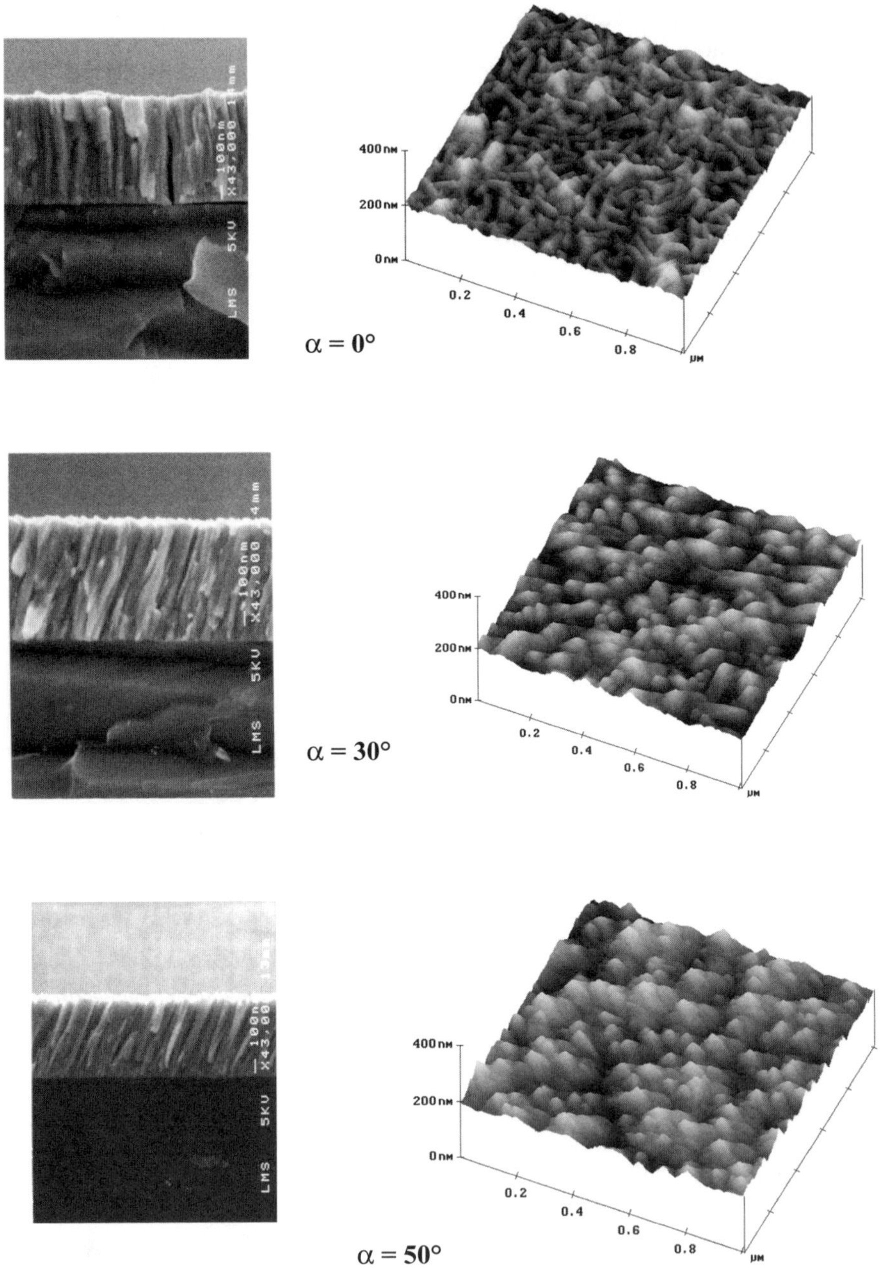

162 Materials and Surface Engineering in Tribology

α = 0°

α = 30°

α = 50°

Figure 3.37. *Microstructure and surface morphology of oblique chromium films as a function of various angles α [LIN 03b]*

Apart from oblique films such as shown in Figure 3.37, it is also possible to deposit zigzag-structured films consisting of successive layers alternately inclined in opposite directions relative to the normal to the surface of the sample. This is shown in Figure 3.38. To achieve this, it is necessary to tilt the sample in two directions relative to the incident particle flux, alternatively (+α and –α). If, in addition to this inclination, the sample is also subjected to a (continuous) rotation in its reference plane (see angle φ in Figures 3.39 and 3.40), it may be possible to generate C-shaped, S-shaped or even helix-shaped films [LIN 03a, SATOM 00, VANP 05].

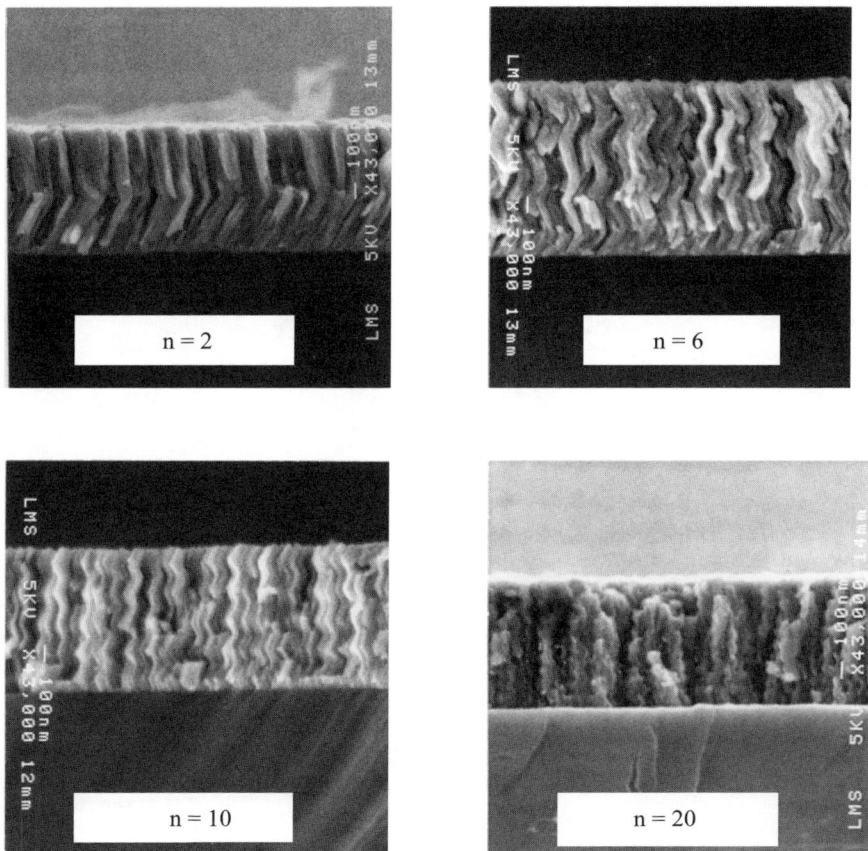

Figure 3.38. *Zigzag-shaped chromium films. The inclination angle α is 50° and n indicates the number of deposited layers in each case [LIN 04]*

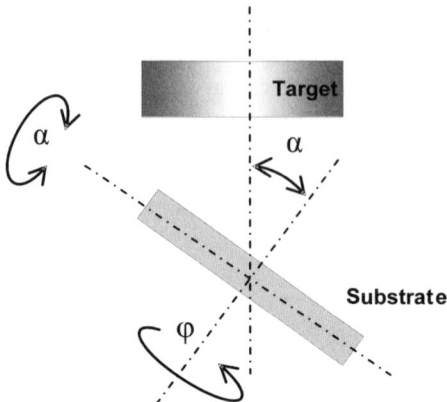

Figure 3.39. *Schematic representation showing the movements of a sample-holder that allows the deposition of successive layers with different microstructures (figure inspired from [LIN 03a])*

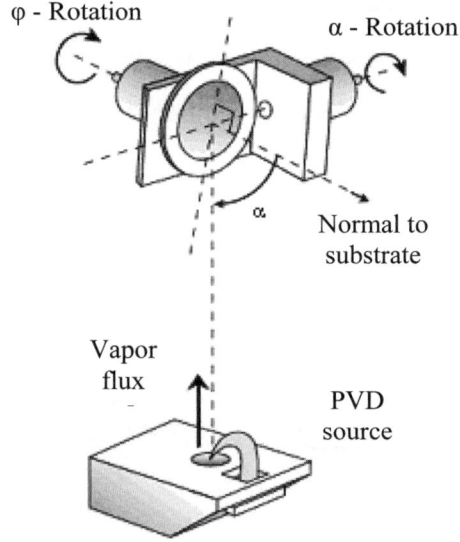

Figure 3.40. *Schematic representation of the principle of controlled deposition using the GLAD technique. Two motors allow the sample to undergo rotations in two orthogonal planes, thus allowing the desired structure to be obtained (see Figure 3.4) [VANP 04]*

Figure 3.41 shows an example of a multilayered coating produced using several different materials and comprising several kinds of microstructures [VANP 04].

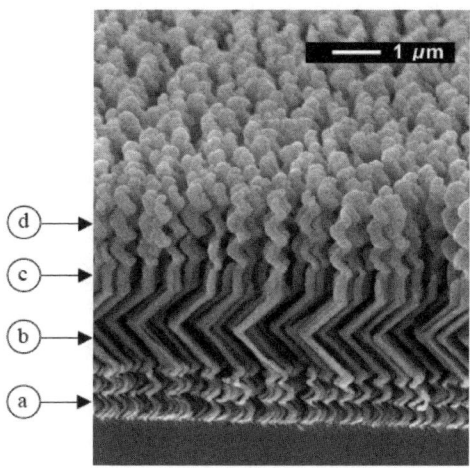

Figure 3.41. *Composite multilayered film showing the possibilities of the GLAD technique [VANP 04]. Layer a: TiO_2 right-handed helix, 450 nm step, 3 turns; layer b: SiO_2 zig-zag 1650 nm thick; layer c: SiO_2 vertical posts 600 nm thick; layer d: SiO_2 left-handed helix 580 nm step, 3 turns*

These GLAD-generated films represent a new class of materials suited to multiple applications. Indeed, due to the possibility to engineer the surface of a material through the deposition of structured films (zigzags, helixes, columns, etc.) of controlled porosity, GLAD opens up opportunities for very diverse applications such as humidity or gas sensors. Indeed, it has been shown that sensors based on such topographically structured materials exhibit improved performance, particularly with regard to their shorter response times compared to conventional sensors. Other applications in the field of photonics, energy storage, display technology and magnetic data storage have also demonstrated the wide applicability of this technique [SATOM 00].

As a consequence of their specific structural properties, these materials are expected to become increasingly important for tribological applications. An unfortunate result of their very recent emergence, however, is that only a limited number of publications have explored this domain. We highlight here three significant studies. In the first study [SET 99], the authors used nano-indentation to study the mechanical behavior of helicoidal SiO films, and showed that the films exhibit a genuine spring effect when deformed (Figure 3.42). Two years later, the same group published additional results concerning the mechanical characterization of films generated using the GLAD technique [SET 01]. More recently, another group [LIN 06] modeled the behavior of zigzag-structured chromium layers under nano-indentation. The calculated hardness, toughness and Young's modulus were found to be in good agreement with experimental measurements.

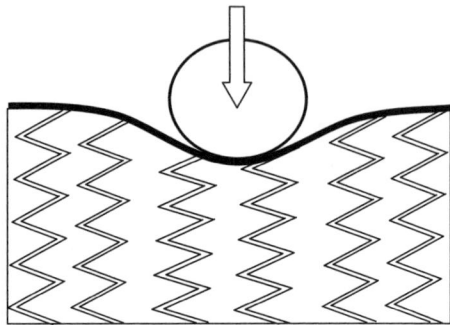

Figure 3.42. *Elastic deformation or "spring effect" of helix or zigzag-shaped films subjected to nano-indentation (based on experimental results [SET 99])*

Apart from the spring effect illustrated in Figure 3.42, the potential of these materials to yield controlled roughness and porosity means that materials perfectly adapted to lubricated friction can be produced. Moreover, oblique columnar structures should enable the development of surfaces presenting anisotropic friction characteristics. Indeed, Figure 3.37 clearly shows that if a sphere is placed under friction against an oblique columnar film, the friction will vary according to the sliding direction of the sphere relative to the angle of the columns.

3.4. Hard anti-wear and decorative coatings

Vapor deposited hard coatings are mainly used as anti-wear materials, but they can also be used in decorative applications when they combine low abrasive wear and attractive colors.

3.4.1. *Hard anti-wear coatings*

3.4.1.1. *Transition metal nitrides*

Simple or mixed nitrides such as TiN, AlN, TiAlN, CrN or ZrN have been extensively studied, and some are used more specifically as anti-wear coatings/materials. They are deposited using the PVD or PACVD techniques in the form of a layer with thickness ranging from one to five microns.

Titanium nitride (TiN) is the most commonly used of these materials. From the crystallographic point of view, it is characterized by a face-centred cubic lattice as shown in Figure 3.43. TiN hardness can range from 1400 HV to 4000 HV depending on the deposition technique and conditions [TAK 97a].

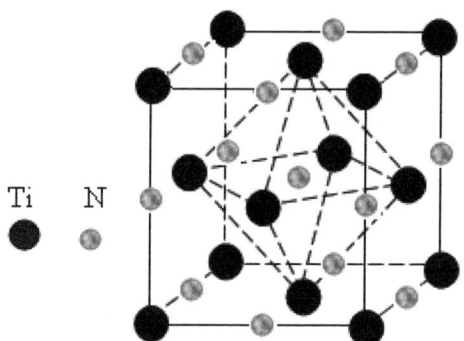

Figure 3.43. *The face-centered cubic lattice of titanium nitride*

Titanium nitride readily oxidizes at 500°C yielding TiO_2, and this significantly impairs its mechanical properties. This material decomposition, under conditions of extreme stress, restricts its use. To overcome this, titanium is therefore often used in conjunction with aluminum in the form of a ternary solid solution TiAlN. Due to the formation of the protective oxides Al_2O_3 and TiAlO, this mixed nitride has much better mechanical and tribological properties [RAU 00] and significantly increased resistance to oxidation than TiN [CHU 97, JOS 95].

TiN, TiCN and TiAlN are widely used as coatings for cutting, stamping and metal embossing tools. Because it is chemically more stable, TiAlN allows cutting to be carried out even without lubrication and at greater speeds than with TiN and TiCN.

The addition of silicon also results in significant improvements in the mechanical and thermal properties of TiN. The compound TiSiN produced through PECVD is a nanocomposite material made of TiN nanoparticles in an amorphous matrix of SiN_x. It is characterized by good hardness (50–70 GPa), excellent resistance to oxidation and, even with relatively low levels of silicon (5%), it remains stable up to temperatures of the order of 800°C [DIS 99, LI 92, VEP 95, VEP 00].

The addition of silicon also has beneficial effects on the tribological properties and ability of other nitrides – such as zirconium nitride – to withstand high temperatures [PIL 06]. In addition, silicon results in the formation of a nanocomposite material made of nanocrystals less than 10 nm in size.

The characterization of the tribological properties of ZrSiN as a function of the quantity of added silicon has shown that the friction coefficient (see Figure 3.44) and rate of coating wear drop sharply when the silicon content reaches 7.6% (see Figure 3.45a and 3.45b) [PIL 06]. This can be partly accounted for by the amorphization of the material, partly by the oxidation of silicon under friction and by the generation of a mixed film made of hydrated silicon oxide and hydroxide, which protects the surface and behaves as a lubricant [BERTR 00, LIUC 01, SATOT 94].

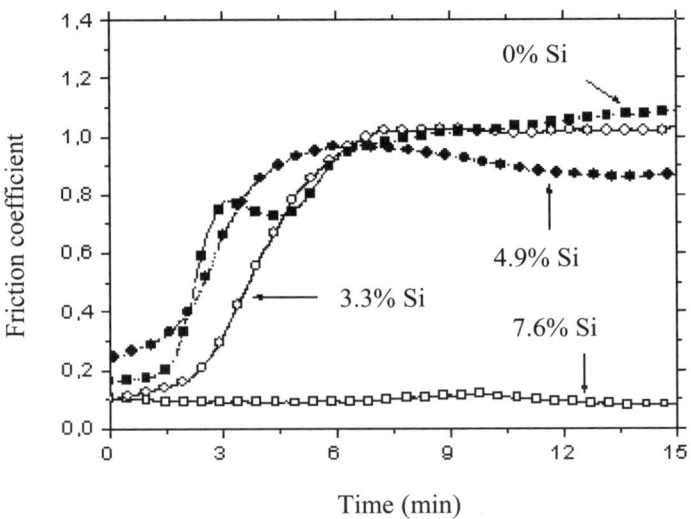

Figure 3.44. *Variation in the friction coefficient as a function of time for a PVD film of ZrSiN (using a ball-on-disc tribometer). The sphere is a 5 mm diameter alumina ball, the load is 25 N and the sliding speed is 5 mm s^{-1} [PIL 06]*

Materials for Tribology 169

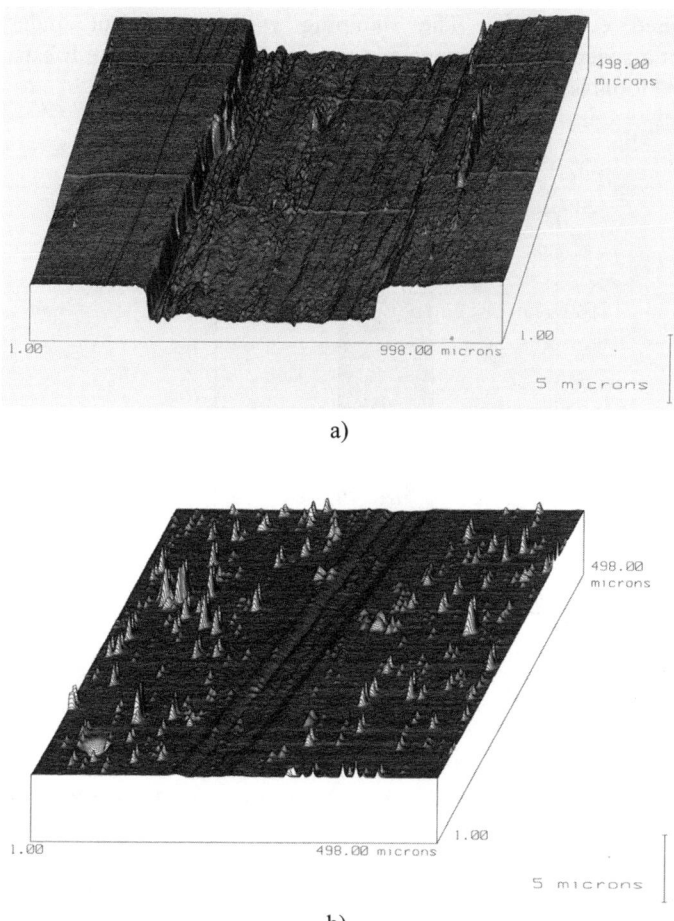

Figure 3.45. *Wear path at the end of the friction test described in Figure 3.44 for: a) ZrSiN containing 3.3% silicon (similar results were obtained for a silicon content of 0 and 4.9%); b) ZrSiN containing 7.6% silicon [PIL 06]*

Chromium nitride is characterized by greater thermal stability [NAV 93] and better resistance to wear [HUA 94, SATOT 94, YAO 97] and corrosion [BERTR 00, LIUC 01] than titanium nitride. As a result, it offers an attractive alternative to TiN for the coating of tools and wear parts.

For example, stamping tools made of X160CrMoV12 steel were coated with CrN or TiN for an industrial application aimed at the production of copper contacts for electrical components. These were then subsequently tested and compared to their untreated equivalents. The stamping was carried out under lubricated conditions at a speed of 350 stamps per minute. The comparative lifespan of treated and untreated tools is presented in Figure 3.46.

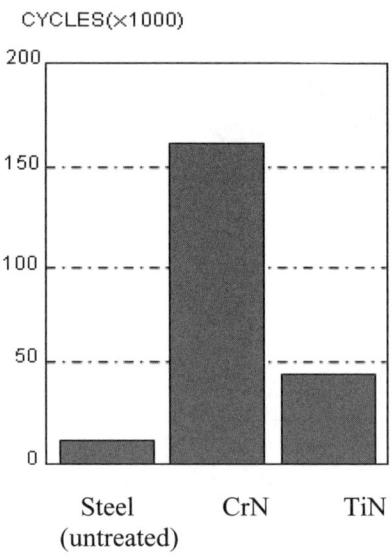

Figure 3.46. *Comparative lifespan of X160CrMoV12 steel stamping tools when untreated, coated with CrN or coated with TiN [TER 96]*

The lifespan of CrN-coated stamping tools is 14 times greater than that of their untreated equivalents and 3.5 times greater than those coated with TiN. Moreover, copper has been shown to adhere significantly to TiN coatings and to induce a significant transfer of metallic material unto the ceramic. However, in the case of CrN coatings, the transfer of material and the friction coefficient proved extremely low (0.25 for CrN but 0.35 for TiN) [TER 96].

When deposited using reactive sputtering, CrN can exhibit a wide range of mechanical and microstructural characteristics, depending on specific deposition conditions [HAE 02]. In particular, the bias voltage of the sample plays a very important part. For example, the hardness of a chromium nitride film measured at 700 HV for an unbiased sample was increased to 2100 HV with a –150 V applied bias. When measured using wear tests, the coating lifespan was found to increase by a factor of ten using a substrate biased to –225, –300 or –900 V when compared to deposition on an unbiased substrate.

In the same study [HAE 02], the authors have clearly shown that the total gas pressure and the N_2/Ar pressure ratio plays an equally important part in determining both the microstructure characteristics (grain size, roughness and porosity) and the mechanical and tribological properties of the films. In particular, they demonstrated that film microstructures characterized by high roughness and extensive, deep porosities presented excellent behavior under lubricated friction because the porosities acted as lubricant reservoirs.

3.4.1.2. Carbon-based films

Carbon-based films are an important class of materials for tribology. They are usually generated using the PVD or PECVD techniques.

Diamond-like Carbon or (DLC) generated using PECVD combines high values for hardness (2000–10 000 HV) with a low friction coefficient (0.1–0.2) and great chemical stability [BUL 95, GRI 93, LEH 96, LIUY 96].

DLC is made from a mixture of graphite (sp^2 hybridization state) and diamond (sp^3 hybridization state). Its properties can vary considerably as a function of the graphite/diamond ratio and the proportion of hydrogen it contains. Its mechanical properties can be rapidly degraded when the temperature exceeds 400°C [TAK 03]. In this case, there is a sharp drop in the coating hardness due to the dehydrogenation of the material and the transformation of the carbon atoms from the sp^3 to the sp^2 form.

The tribological behavior of DLC is also very sensitive to humidity, but published results in the field are often contradictory [AKI 04, BUL 95, GRI 93, LEH 96, LIUY 96, TAK 03]. Some publications suggest that higher rates of humidity trigger a reduction in the friction coefficient and an improvement in wear resistance, but many others argue that the opposite occurs.

These contradictory results are partly due to differences in the composition and microstructure of the different films analyzed. Indeed, there is an important volume of hydrogen contained in the films, the graphite/diamond ratio and the degree of purity and density of the films. These data are not always provided in sufficient detail in the published material.

DLC makes an excellent anti-wear coating, and is particularly well-suited for cutting tools used in the machining of copper-aluminum alloys. However, because of a strong reaction between the carbon and iron, it cannot be used for the machining of steels.

DLC is also often used to coat the moulds required for plastic or aluminum injection molding. Moreover, as it is bio-compatible, it is also the preferred material coating for the surfaces subject to friction in hip and knee prostheses.

Another material, carbon nitride (or CN_x), is also worth mentioning for its superior mechanical and tribological properties. Indeed, these properties are similar to – if not better than – those of DLC, but CN_x can also withstand high temperatures without degradation [TAK 03].

Carbon nitride films have been the subject of much experimental research as a result of theoretical calculations that showed that the hardness of the crystalline compound C_3N_4 was equivalent to, if not greater than, that of diamond [LIUA 89, LIUA 90]. However, the synthesis of this material remains the subject of some controversy because its method of production is not yet fully understood or reproducible. However, it has been possible to generate and characterize numerous CN_x compounds that do not have the stoichiometry of the C_3N_4 compound. It has also been shown that carbon nitride films deposited by sputtering and containing 10% nitrogen had better wear resistance and a lower friction coefficient than pure carbon films generated under the same conditions [KHU 96].

Another study has shown that the hardness of films generated using the sputtering technique could increase from a value of 9 GPa with pure carbon coatings to 25 GPa for carbon nitride films with an 18% nitrogen concentration [CUT 96]. However, many published results argue that high rates of nitrogen tend to reduce material hardness due to the increase of sp and sp^2 carbon at the expense of sp^3 carbon [DEGR 98].

An increase in the friction coefficient and an improvement in resistance to wear have also been observed with an increase in the quantity of nitrogen [KUS 98].

Another noteworthy carbon-based coating is the WC/C compound produced through PVD with hardness ranging from 1500 to 2000 HV. By alternating hard tungsten carbide with lubricating carbon layers, it is possible to obtain a high-performance multilayered material well-adapted to the coating of tools used in the machining and embossing of aluminum and metallic sheets with high Young's modulus [POD 04, WANS 99]. This coating can be applied directly onto the tools or deposited as a final coating after prior treatment of the substrate with an initial film of TiN, which enhances the adhesion of the WC/C coating to the substrate and also acts as a diffusion barrier. Other nitride films such as CrN can also be deposited in this way.

3.4.1.3. *The role of the substrate*

Hard coatings should always be applied onto hard substrates to guarantee maximum efficiency.

A useful guideline is that a hard coating of 1500 HV should never be applied onto a substrate with hardness lower than 700 HV. Indeed, softer substrates subjected to friction will undergo plastic deformation and thus be strained to the extent that the hard but brittle coating will crack and peel off from the substrate (see Figure 3.47).

This has been clearly shown in a comparative study of three distinct layers of TiN (2100 HV), TiCN (2600 HV) and DLC (3800 HV) deposited onto two substrates of different hardness: 35CrMo4 steel of hardness 250 HV and X85WMoCrV06.05.04.02 steel of hardness 880 HV [TAK 97b].

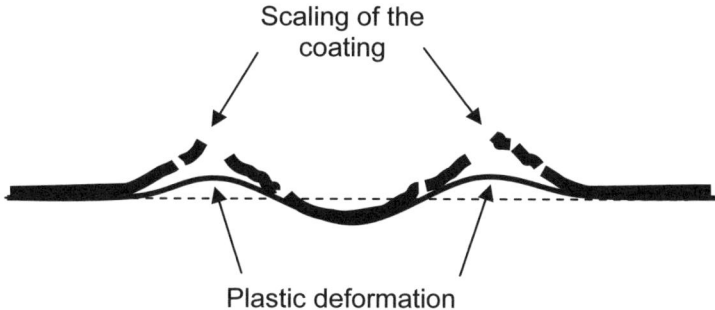

Figure 3.47. *Behavior of a ceramic layer deposited onto a steel substrate subject to friction. When the substrate is soft, it deforms plastically and causes the coating to peel off. The minimum hardness required for the substrate is about 700 HV*

In order to optimize the properties of hard coatings, multilayered deposits of graded hardness such as Ti/TiN/TCN/DLC, Ti/TiN/WC/C or TiN/TiCN/CN [MOR 04] have provided excellent results under friction.

More complex systems with composite films also exhibit high performance. This is, for example, the case with the multilayered coating Ti/TiN/TiCN/TiC/(Ti-DLC)/TiC/(Ti-DLC) in which the titanium film improves the adhesion of the coating to the steel substrate, while the intermediate layers support the load and act as diffusion barriers or crack breakers. The outer DLC layer in this case is designed to lubricate the contact by reducing the friction coefficient [VOE 96] (see Figure 3.48).

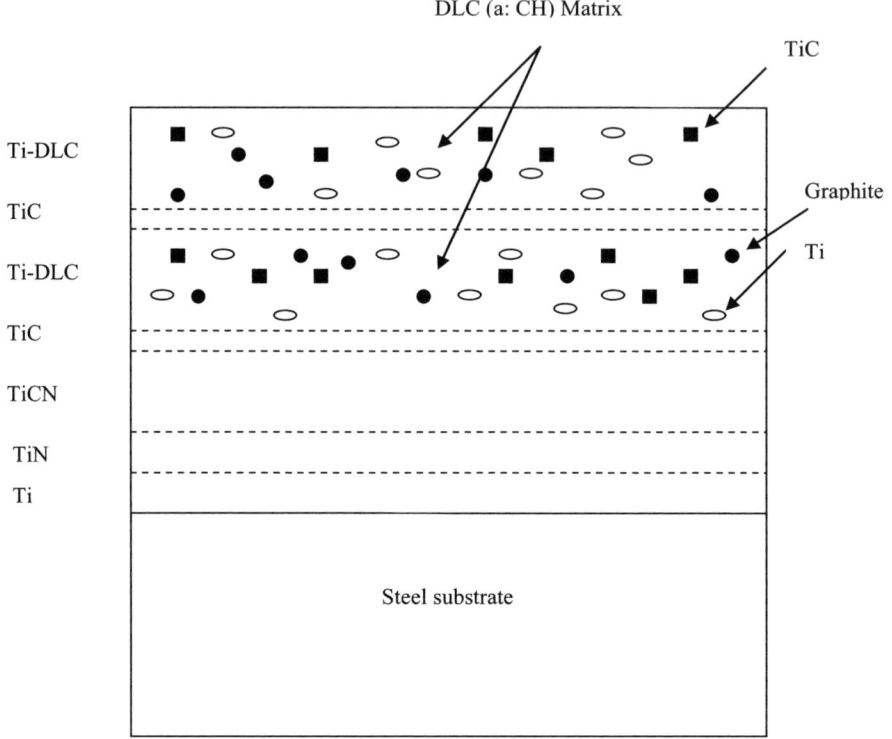

Figure 3.48. *Example of a multilayered coating [VOE 96]*

3.4.2. Decorative coatings

The oxide, nitride, oxinitride, oxicarbide or carbonitride coatings of certain elements such as Ti, Zr, Cr or Al allow layers of different colors to be produced and are therefore particularly well-suited for decorative purposes. The manufacturing of spectacle frames, jewellery, watches and cutlery provide numerous examples of such applications.

Film and coating color is described using the CIE system introduced by the International Lighting Committee (*Comité International de l'Eclairage*). With this system, color is expressed using three parameters or coordinates L*, a* and b*. The coordinate L*, which expresses luminosity, ranges from 0 for a black object to 100 for a white object and varies continuously between these extremes. L* is represented as a vertical axis, whereas the horizontal plane comprises the two axes a* and b*. The coordinate a* indicates position between red and green (positive values of a* indicate red while negative values of a* indicate green), whereas the coordinate b* indicates position between yellow and blue (positive values of b* indicate yellow while negative values of b* indicate blue) (see Figure 3.49).

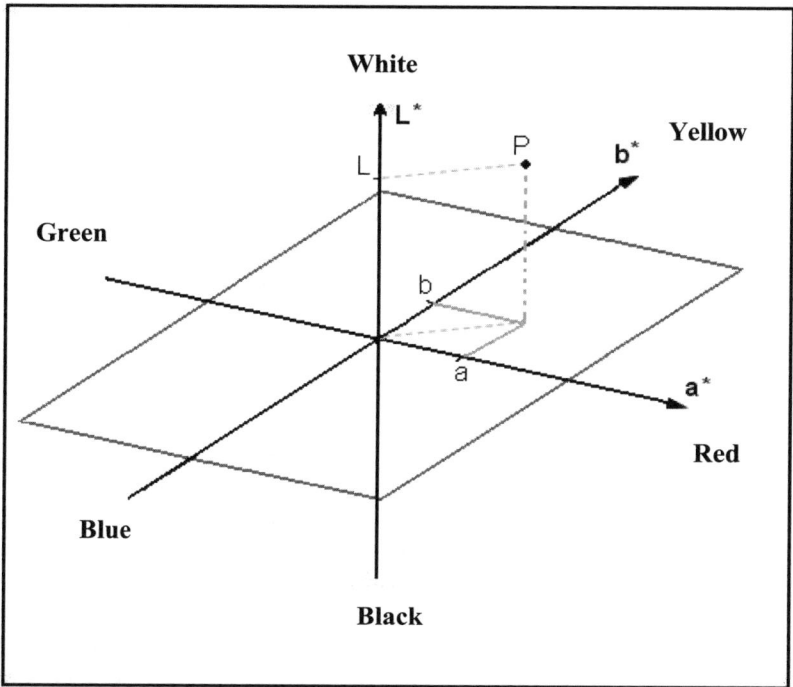

Figure 3.49. *Representation of a point P at coordinates L, a, and b in the color space L*a*b**

We limit ourselves here to coatings produced using PVD (generally applied using reactive cathodic sputtering) as this technique is the most widely used in industry within the decorative sector.

The substrates that can be used with this deposition technique are those that can withstand temperatures of the order 200°C without degassing.

Among the coating materials that are used, titanium dioxide (TiO_2) has the advantage of generating so-called interferential films where the color varies as a function of the film thickness. For example, it is possible to obtain white films for coating thicknesses less than a few nanometers, yellow films when the thickness exceeds 40 nm, red films for thicknesses around 50 nm, and blue and green films for thicknesses between 50 and 100 nm [MANA 04].

The fact that the films are based on optical interference, however, has limited the widespread use of TiO_2 coatings on an industrial scale. In fact, only flat surfaces can be coated satisfactorily using this technique and, for complex surface geometries, different parts of the substrate are not coated with the same thickness, leading to differences in colors. Moreover, titanium dioxide films are brittle and their hardness is limited; the combined interferential nature of the coating and its low abrasion resistance therefore limits the use of this material in decorative applications.

In contrast to TiO_2, titanium nitride (TiN) is a hard, non-interferential coating which presents good resistance to abrasion and appears golden yellow. Its remarkable mechanical properties and its attractive color mean that it is one of the materials most widely used for decorative purposes.

However, the radiance and golden yellow color of TiN are not rigorously identical to those of gold and, in spite of systematic work carried out on the production of TiN films under varying experimental conditions, it has not been possible to achieve a genuine golden yellow color [MANA 04].

There is no possible combination of the three parameters a*, b* and L* for the different layers generated that corresponds totally to the reference parameters of the gold standards (according to the normalized NIHS colors or *Normes de l'Industrie Horlogère Suisse*). Indeed, even when two parameters coincide with the characteristics of one of the gold standards, the third is very far from the expected value. It is for this reason that in industrial applications such as watch-making, jewellery and accessories, the 1 μm thick film of TiN is coated with a additional layer of "gold flash" one-tenth of a micron thick.

This solution, widely used in industry, does however present a major drawback. In particular, the thin layer of gold (a soft material) erodes relatively quickly and reveals the slightly darker titanium nitride layer, which causes the surface to look tarnished [CONSTANTI 06]. To solve this problem, most research in the field has been aimed at developing hard, genuinely gold-like films. Among some of the solutions envisaged are mixed TiN/ZrN films, TiZrN films or multilayered TiN/ZrN coatings.

These multilayered TiN/ZrN coatings appear to open up interesting opportunities. Indeed, alternate nanometer-scale films of TiN/ZrN deposited using reactive cathodic sputtering have managed to reproduce the gold standards 1N14, 2N18 and 3N18 by combining TiN (golden-yellow but too reddish) and ZrN (golden-yellow, but too greenish). Optimizing the experimental process has yielded the same color values as the 1N and 2N gold standards for the a* and b* values. However, the third value L* requires further improvement [CONSTANTI 06, MANA 04].

Between the hard, non-interferential golden-yellow titanium nitride and the softer, interferential titanium dioxide, the class of titanium oxinitrides $Ti_xN_yO_z$ presents intermediate properties that yield hard, non-interferential coatings presenting desirable colors for decorative applications. Indeed, depending on the oxygen/nitrogen ratio, titanium oxinitrides can take on a wide range of colors [MARTIN 02, VAZ 04].

$Ti_xN_yO_z$ films have recently been generated using the reactive sputtering technique with pulsed gases [CHAP 05, MARTIN 02]. Many ternary compounds ranging from TiN to TiO_2 were thus obtained and characterized, showing that the shift from nitride to oxide is paralleled by a gradual decrease in hardness, the Young's modulus and conductivity.

Regarding the colors obtained, the films are interferential within the composition range (between 30 and 100% oxygen); however, when oxygen content is between 0 and 30%, they are non-interferential and present much higher values for hardness than TiO_2.

Increasing the oxygen content has yielded a varied palette of colors. It has been possible to continuously vary the color from golden yellow for TiN to metallic bronze, and then blue, purple, and dark and light green for compounds higher in oxygen content.

Red is a rare and highly sought-after color in decorative applications. It is therefore significant that recent research has reported that the Fe-O-N system has made it possible to obtain compounds presenting the same reddish color as hematite (α-Fe_2O_3). Iron-oxinitride films were then generated using magnetron sputtering by simultaneously injecting oxygen and nitrogen into the experimental reactor [PET 06].

Carbonitrides are yet another class of coatings which must be mentioned as they offer interesting solutions for decorative applications. Indeed, these materials are characterized by high mechanical properties and highly sought-after colors. The principal examples are titanium carbonitride (bronze, gray), titanium/aluminum carbonitride (black, gray and aubergine), chromium carbonitride (gray) and zirconia carbonitride (brass yellow) [CAT 05].

3.5. Characterization of coatings: hardness, adherence and internal stresses

3.5.1. *Hardness*

We define three degrees of hardness principally differentiated by the depth of indentation (d) as summarized in Table 3.6. The load required to reach these depths will vary according to the material being tested. With nano-hardness, the required load ranges from a few micronewtons to several hundreds of millinewtons. With micro-hardness, the required load ranges from a few tens of millinewtons to a few newtons. Finally, with macro-hardness the required load ranges from a few newtons to several tens of newtons.

	Nano-hardness	Micro-hardness	Macro-hardness
Indentation depth (µm)	0.001–1	1–50	50–1000

Table 3.6. *The three types of hardness*

As a consequence, the choice of hardness test will be tailored to the type of material under study as well as to the application envisaged. Nano-hardness will therefore be preferred to characterize thin layers (less than a micron) while micro-hardness will be used to analyze the effects of surface treatments such as shot peening or thermochemical treatments (carburizing, nitridation, etc.). Macro-hardness will typically be used to measure the hardness of steels following treatments such as quenching or tempering.

In section 1.2.4.1, we described the basic principles of the hardness test and the different methods used to measure the hardness of materials. In the case of coated materials, carrying out hardness tests presents two major difficulties. Specifically, measurements carried out under very light loads can be affected by a size effect due to the close proximity of the surface (see section 3.5.1.1), while measurements performed under very heavy loads can be affected by the proximity of the substrate (see section 3.5.1.2).

3.5.1.1. *The indentation size effect*

Hardness is a mechanical property of the material and its value (determined experimentally) should therefore not depend on the measurement test conditions. However, we note that measured hardness values are systematically higher when the load (and penetration depth) is lower. Many reasons have been suggested to account for this indentation size effect (ISE): work hardening; the presence of a superficial oxide layer; the influence of impurities segregating on the surface after annealing of the material; the presence of a free surface acting as an obstacle to dislocation movements; or even elastic return. Indeed, although elastic return is often negligible in macro- and micro-hardness, it becomes significant when the impressions obtained present indentation depths ranging from a few tens to a few hundreds of nanometers [BUL 01a]. Another theory based on the concept of geometrically necessary dislocations (GND) to accommodate the deformation gradient in the plastified zone has also been suggested [NIX 98].

Two expressions are generally used to characterize the influence of the ISE on the hardness:

– Model 1 [NIX 98]

$$\frac{H}{H_0} = \sqrt{1 + \frac{d^*}{d}} \qquad [3.38]$$

where H is the measured hardness, H_0 is the measured hardness for an indentation depth large enough so that ISE is not present, d is the indentation depth and d* is a characteristic length that depends on the shape of the indenter determined from the fitting of the experimental points by the model.

– Model 2 [BUL 01a]

$$H = H_0 + \frac{k}{d} \qquad [3.39]$$

where H is the measured hardness, H_0 is the measured hardness for an indentation depth large enough so that ISE is not present, d is the indentation depth and k is a constant determined from the fitting of the experimental points by the model as shown in Figure 3.50. It clearly shows the validity of the model and illustrates the significant influence of elastic return on the ISE. Indeed, nano-indentation tests carried out using a blunt-tip indenter clearly yielded greater elastic return than those performed with a sharp-tip indenter.

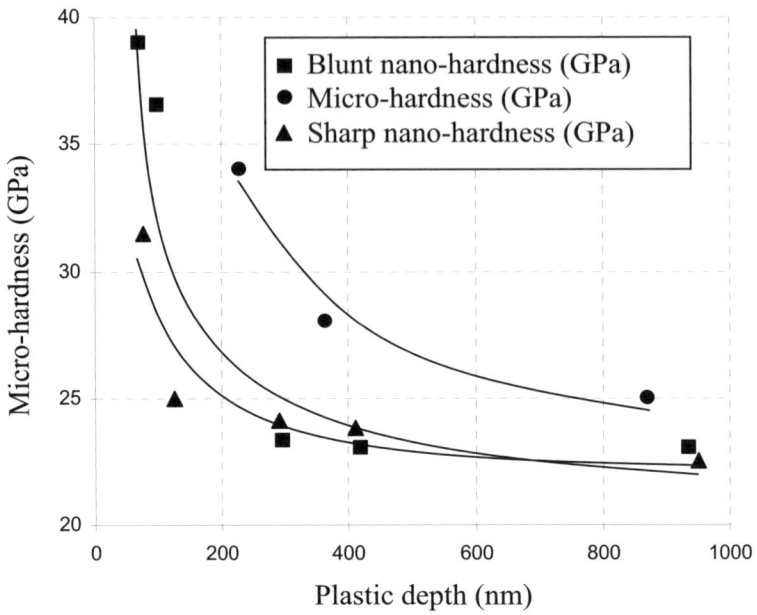

Figure 3.50. *Hardness variation with plastic depth for PVD TiN, comparing experimental data (symbols) with model fits (solid lines). The indentation size effect behavior is clearly visible and depends on indenter geometry and data analysis method [BUL 01a]*

3.5.1.2. *Hardness tests for coated materials*

The experimental determination of the mechanical and tribological properties of coatings is often a delicate procedure due to the proximity of the substrate which can significantly affect measurement results.

In the case of hardness tests, if h and d represent the thickness of the coating and the depth of penetration of the indenter into the surface (see Figure 3.51), then as the ratio d/h increases beyond a critical value $(d/h)_c$ the measurements will be affected by the proximity to the substrate. The hardness measured will therefore be overestimated or underestimated depending on whether the hardness of the substrate is higher or lower than that of the coating (see Figure 3.52).

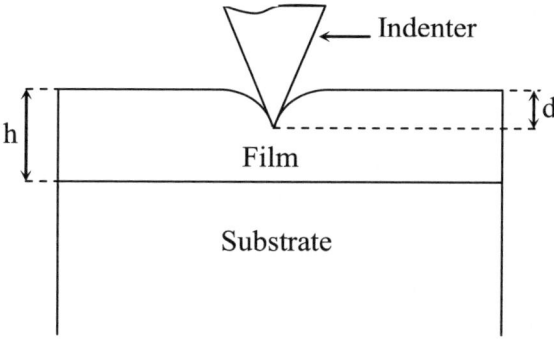

Figure 3.51. *Indentation of a coated material*

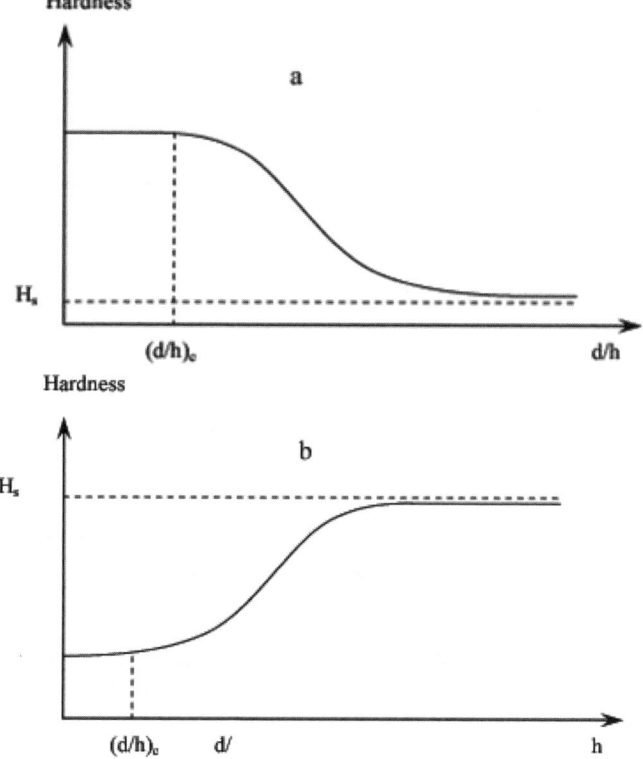

Figure 3.52. *Hardness variation with the d/h ratio (see Figure 3.51). H_s is the hardness of the substrate: a) the coating is harder than the substrate; b) the coating is softer than the substrate. The indentation size effect has been neglected*

Many models have been designed to analyze hardness tests performed on coated materials, as described in the following.

3.5.1.2.1. Buckle's model [BUC 65]

Using experimental data, Buckle was the first to suggest a model allowing the influence of the substrate to be taken into account. He divided the volume under stress into twelve layers of equal thickness d parallel to the surface and attributed to each layer a relative weight P_i and hardness H_i (see Figure 3.53). The contribution of a given layer i to the total hardness measured will be given by the product H_iP_i. The layer thickness is related to the indentation depth by the relation $h = nd$, where n is an integer corresponding to the order (i) of the last layer d which still forms part of the coating, with the zone $i = n+1$ then belonging to the substrate.

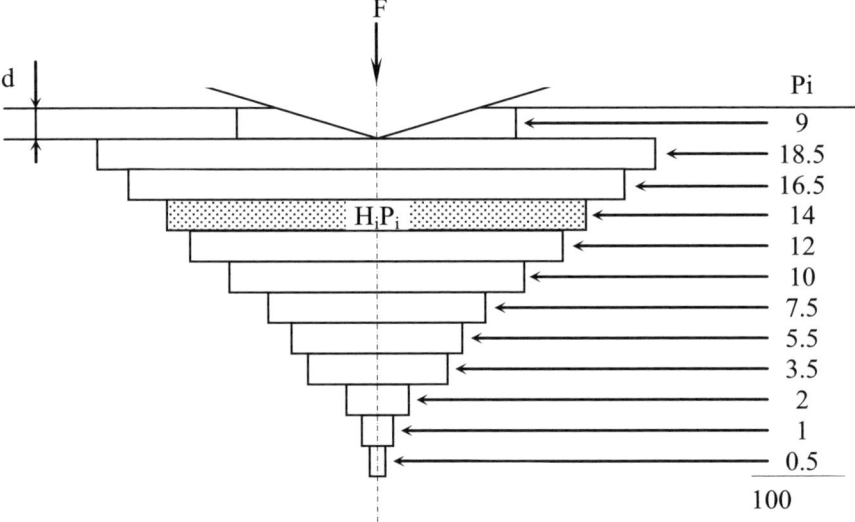

Figure 3.53. *Buckle's model of indentation. The volume under stress is divided into twelve layers of equal thickness d parallel to the surface [BUC 65]*

If the hardness of the substrate is given by H_s and if H_f is the hardness of the film, Buckle's model allows the hardness of the coated material (noted H_c and hereafter referred to as composite) to be expressed:

$$H_{c(d)} = \frac{\sum H_i P_i}{\sum P_i} = \frac{H_f \sum_{i=1}^{n} P_i + H_s \sum_{i=n+1}^{12} P_i}{\sum_{i=1}^{12} P_i} \quad [3.40]$$

This equation gives the hardness of the composite for an indentation depth d.

If we write:

$$a = \frac{\sum_{i=1}^{n} P_i}{\sum_{i=1}^{12} P_i} \quad \text{and} \quad b = \frac{\sum_{i=n+1}^{12} P_i}{\sum_{i=1}^{12} P_i} \quad [3.41]$$

then equation [3.40] may be written:

$$H_{c(d)} = aH_f + bH_s \quad \text{with} \quad a+b = 1 \quad [3.42]$$

If n is known, equations [3.41] can be used to calculate the values of the coefficients a and b for different values of d or d/h. Figure 3.54 shows the variation of a as a function of the d/h ratio. It is interesting to note that if the d/h ratio is lower than 0.1, the measurement is only weakly influenced by the proximity of the substrate. This observation forms the basis of the so-called one-tenth rule of thumb that states that hardness measurements of a coated material can be deemed reliable and unaffected by the substrate provided the depth of penetration of the indenter does not exceed one-tenth of the coating thickness. This rule, established experimentally based on Vickers hardness tests, is generally valid for this type of hardness but only in the case of hard coatings deposited onto soft substrates. When the coating is softer than the substrate, a significant pill-up forms around the indentation due to the plastification of the soft film (see Figure 3.55) and this considerably affects the predictions of the model. Indeed, this pill-up partially withstands the applied load as reported in [CHA 88] and confirmed in [IOS 96].

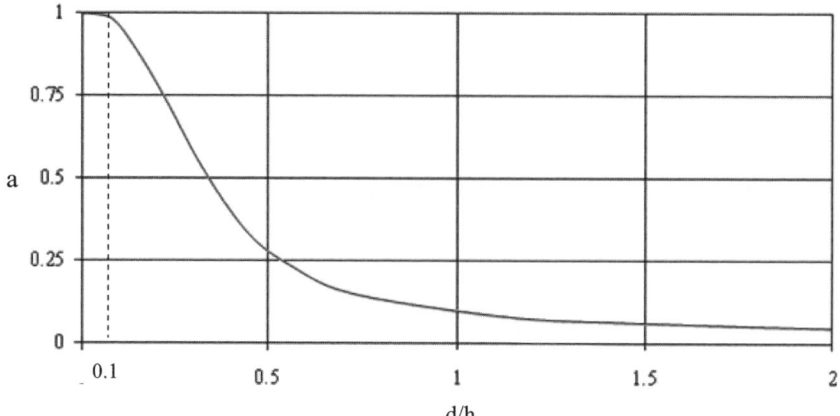

Figure 3.54. *Variation of a (see equation [3.41]) with d/h ratio (see Figure 3.51)*

Generally speaking, publications report values for the $(d/h)_c$ ratio ranging between 0.07 and 1 [BEE 05, CHE 95, IOS 05, KOR 98]. Table 3.7 presents some results obtained with different coatings. These results confirm that the one-tenth rule is generally true in those cases when the coating is harder than the substrate. However, significant variations are noted in the reverse situation when the substrate is harder than the coating.

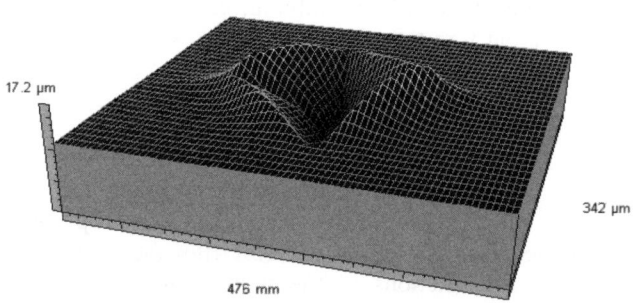

Figure 3.55. *Indentation of a nickel film (hardness = 250 HV) deposited onto a harder steel substrate of hardness 700 HV; a pill-up forms around the indentation*

Reference publication	Substrate	Coating	Relative hardness ($H_{coating}/H_{substrate}$)	Critical depth $(d/h)_c$
[HUM 83]	Stainless Steel	TiC	20	0.10
		TiN	15	0.14
[TAZ 78]	Glass	Cu, Ag, Au	<0.10*	1
[BAN 85]	Silicon	Al	<0.10*	0.20–0.40
[LEB 85]	Quenched steel (4135 (SAE))	Cu	0.30	0.22
	Tempered steel (4135 (SAE))	Cu	0.50	0.29
	Quenched steel (4135 (SAE))	NiB	1.30	0.15
	Steel (1018 (SAE))	NiB	6.40	0.09
[KOR 98]	Stainless steel M304	CrN	10	0.10
	Tool steel ASP23	TiN	4.33	0.10
[BEE 05]	Silicon	Cu	0.17	0.30
[CHE 95]	Aluminum	Al_2O_3	36	0.10–0.15
	Silicon	Al_2O_3	0.75	0.40–0.50
	Sapphire	Al_2O_3	0.36	0.70–0.90
[DUR 91]	Copper	Ni	2.08	0.12
	Iron	Ni	1.40	0.16
	Steel C48	Ni	0.56	0.33
	Quenched steel C48	Ni	0.25	0.47

Table 3.7. *Values for the critical ratio $(d/h)_c$ for some coating/substrate pairs (*estimated values)*

3.5.1.2.2. The Jonsson and Hogmark model [JON 84]

This model is expressed in terms of the ratios of the surface areas of Vickers indentations made on the film and substrate, and gives the composite hardness as:

$$HV_c = \frac{A_f}{A} HV_f + \frac{A_s}{A} HV_s \qquad [3.43]$$

where $A = A_f + A_s$, the sum of the areas onto which HV_f and HV_s hardnesses are applied, respectively (see Figure 3.56).

The hardness of the coating can be expressed as:

$$HV_f = HV_s + \frac{HV_c - HV_s}{2C\dfrac{h}{D} - C^2\left(\dfrac{h}{D}\right)^2} \qquad [3.44]$$

where D is the diagonal of the indentation, h is the thickness of the coating and C is a geometric constant that takes the value 0.5 if the coating undergoes plastic deformation and 1 if the coating deformation is of the brittle type.

This model generally yields reliable results when the depth of indentation is significant (d/h > 1) and the deformation is of the brittle type. However, for less significant depths of indentation, greater variations are noted between experimental values and the predictions of the model [KOR 98].

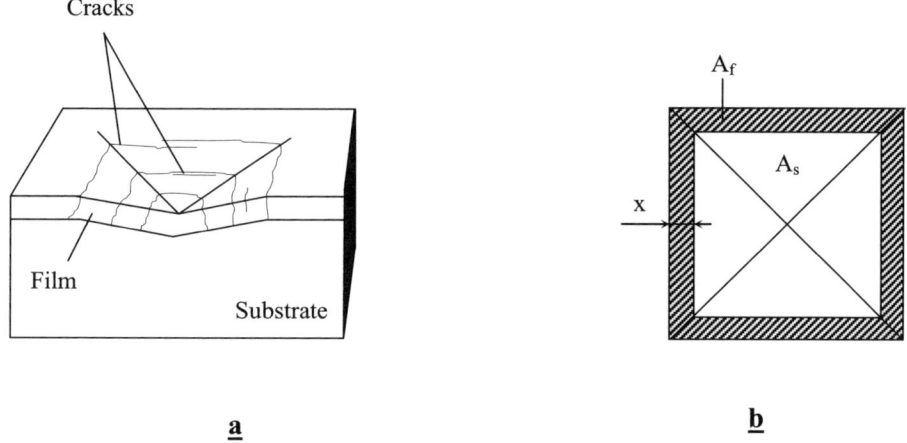

Figure 3.56. *The Jonsson and Hogmark model. Indentation of a film showing (a) a brittle-type deformation and (b) the definition of areas A_s and A_f [JON 84].*

A more realistic approach to describe the hardness of a coated material involves considering the volume resulting from the deformation rather than the surface area subjected to this deformation. This is the approach used in the following models.

3.5.1.2.3. The Burnett and Rickerby model [BUL 01a ,BUR 87a]

This model is an improvement on the Sargent model [SAR 79] which suggests that the composite hardness can be expressed using the following equation:

$$H_c = \frac{V_s}{V}H_s + \frac{V_f}{V}H_f \qquad [3.45]$$

where V ($V_s + V_f$) is the total volume that undergoes plastic deformation between the indenter and a hemisphere with a diameter equal to the indentation diagonal. V_f and V_s are the part of V contained in the film and the substrate respectively (see Figure 3.57a). H_c, H_s and H_f are the hardness values for the composite, the substrate and the film.

When a Vickers indenter is used, the volume V can easily be calculated from the expression suggested by Lawn et al. [LAW 80]:

$$R = \frac{D}{2}\left(\frac{E}{H}\right)^{1/2} \cotan^{1/3}\xi \qquad [3.46]$$

where R is the radius of the plastically deformed zone, ξ is the half angle at the summit of the indenter between two opposite vertices (74° for a Vickers indenter), D is the diagonal of the indentation and E and H are the Young's modulus and the hardness of the material being tested, respectively.

When the film thickness is known, calculating V allows the volumes V_s and V_f to be determined. If the substrate hardness is also known, the experimental determination of H_c can then yield the coating hardness H_f using equation [3.45].

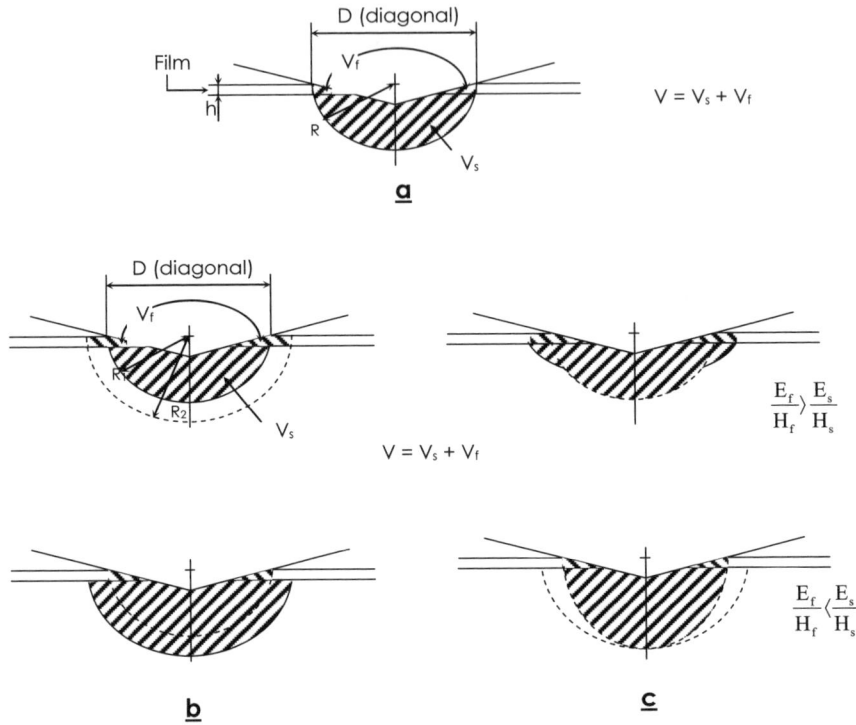

Figure 3.57. *Schematic representation of deformations induced by indentation; a) the volumes considered in the case of the Sargeant model; expected modifications to plastic zone morphology when b) there is no adhesion between film and substrate; and c) when adhesion is strong. E_f and E_s are the Young's moduli for the film and substrate. H_f and H_s are the film and substrate hardness [BUL 01b, BUR 87b]*

Burnett and Rickerby have improved this model by modifying the definition of the volumes V_f and V_s (see Figures 3.57b and 3.57c) to take into account the variation between the shape of the true deformation relative to the geometrically-ideal spherical shape. Considering two hemispheres of radii R_1 and R_2 on either side of the film-substrate interface, the authors have introduced an interfacial parameter χ to take into account the interaction between the coating and the substrate.

The following expressions have been proposed:

– Case 1: $H_s < H_f$

$$H_c = \frac{V_f}{V}H_f + \frac{V_s}{V}\chi^3 H_s \qquad [3.47]$$

where $V = V_f + \chi^3 V_s$;

– Case 2: $H_f < H_s$

$$H_c = \frac{V_s}{V}H_s + \frac{V_f}{V}\chi^3 H_f \qquad [3.48]$$

where $V = \chi^3 V_f + V_s$.

Based on experimental tests carried out on different coatings, the authors have proposed the following expression for χ:

$$\chi = \left(\frac{E_f H_s}{E_s H_f}\right)^n \qquad [3.49]$$

where E_f and E_s are the Young's modulus for the film and the substrate, and n has a value between 1/2 and 1/3.

Other models based on a law of mixtures, including fractional volumes for the coating and the substrate, have also been suggested. Two such models are presented below.

3.5.1.2.4. The Chicot and Lesage model [CHI 95]

This model is based on the superposition of two hypothetical systems representing the volumes resulting from plastic deformation in the film and in the substrate respectively. This model yields the following expression for the composite hardness:

$$H_c = H_s + \left\{\frac{3}{2}\tan^{1/3}\xi\frac{h}{D}\left[\left(\frac{H_f}{E_f}\right)^{1/2} + \left(\frac{H_s}{E_s}\right)^{1/2}\right] - 2\tan\xi\left(\frac{h}{D}\right)^3\left[\left(\frac{H_f}{E_f}\right)^{3/2} + \left(\frac{H_s}{E_s}\right)^{3/2}\right]\right\}$$
$$\times (H_f - H_s)$$

[3.50]

where ξ, D, h, H_f, E_f, H_s and E_s are as previously defined.

When the diagonal of the impression D is greater than the coating thickness h, equation [3.50] is reduced to:

$$H_c = H_s + \left\{\frac{3}{2}\tan^{1/3}\xi\frac{h}{D}\left[\left(\frac{H_f}{E_f}\right)^{1/2} + \left(\frac{H_s}{E_s}\right)^{1/2}\right]\right\}(H_f - H_s) \qquad [3.51]$$

3.5.1.2.5. The Korsunsky *et al.* model [KOR 98]

This model is based on the analysis of the energy required to generate an indentation and yields the equation:

$$H_c = H_s + \frac{H_f - H_s}{1 + k\left(\frac{d}{h}\right)^2} \qquad [3.52]$$

where k is a constant that depends on the film thickness. The other terms have been defined above.

A number of other models have also been suggested, including e.g. the Puchi-Cabrera model [PUC 02], the Bhattacharya and Nix model [BHA 88] or the Lebouvier, Gilormini and Felder model [LEB 85]. Comparative studies of these models are to be found in [IOS 05] and [BEE 05].

3.5.2. *Coating adhesion*

Independently of its mechanical properties, a coating must adhere perfectly to its substrate in order to act as a genuine protective layer against wear. This is why the deposition of protective film coatings is always preceded by some mechanical and/or chemical treatment designed to cleanse the surface and activate it in order to optimize the adherence of the film to the substrate. It is also common to deposit an intermediate film (or binding layer) between the substrate and the final coating. A nickel film is therefore systematically deposited onto copper-based substrates before the application of a gold coating, a silicon film is applied onto steels before they are coated with DLC and a pure titanium film is applied before the deposition of TiN film onto steel substrates.

The very definition of adhesion remains complex and the fundamental mechanisms are varied; they can include mechanical binding, electrostatic forces, diffusion, wetting or chemical bonding [COG 00, DAR 03, ROC 91].

For the case of the combination of two solids A and B having surface energies γ_A and γ_B, the thermodynamic adhesion work or the Dupré adhesion energy is given by the fundamental adhesion relation (see equation [1.13] in section 1.2.3). Adhesion therefore appears as a true material property which needs to be considered in the same way as any other physical constant.

Adherence, also referred to as "practical adhesion", is given by the force or the energy necessary to break the bonds between the coating and the substrate.

Coating detachment never occurs suddenly and completely, but rather arises as a result of the propagation of a crack which gradually breaks the interfacial bonds, liberating elastic energy and allowing the dissipation of irreversible work at the head of the crack [MAU 84]. The crack can propagate when the adhesion energy W is less than the strain energy release rate G (i.e.: W < G). Physically, the quantity G–W represents the driving energy responsible for the crack propagation. W and G can be related by the expression [MAU 78]:

$$G - W = WF(V,T) \qquad [3.53]$$

If the mechanical properties of the materials are known, G can easily be calculated from geometrical considerations such as the size of the pre-existing crack, contact geometry or type of stress.

Equation [3.53] shows that the strain energy release rate depends on two terms: the adhesion energy and a function F(V,T) of the temperature (T) and speed of propagation of the crack (V) which accounts for the viscoelastic losses within the material. F(V,T) is a viscoelastic material property for a given mode of propagation. Adherence will therefore be more significant when the adhesion energy and the viscoelastic losses are high. These losses are particularly significant for polymers and this allows us to "understand why the separation of two glued objects requires such an enormous amount of energy which is at least 10000 times greater than that corresponding to the forces of attraction between molecules" [BAR http].

3.5.2.1. *Methods for adherence testing*

A distinction is usually made between two main classes of techniques: non-destructive and destructive. With non-destructive techniques, an optical or acoustic probe is used to scan the coated material to detect defects or inhomogenities (such as cracks, blemishes, porosity or bubbles) at the film–substrate interface. The most widely used of these methods are acoustic and ultrasound techniques which exploit the fact that material faults or discontinuities modify the propagation speed of acoustic waves or cause wave reflection. Holographic interferometry and infrared thermography are two additional techniques of this type [BAR http].

Destructive techniques can be divided into two categories. In the case of sealed joints or for coatings made of ductile or weakly adhesive materials (such as paints, varnish or polymers), the techniques used can include cross-cutting, cleaving, peeling, blister tests, or three or four point flexion tests. When adherence is high, such as with metallic or ceramic coatings, indentation or scratch techniques are generally used. A selection of these techniques is now presented in more detail [BEN 04, COG 00, DAR 03].

3.5.2.1.1. The peeling test

This test consists of applying a force F at an angle θ to a band of width b deposited onto a substrate, as shown in Figure 3.58. This force is gradually increased with time until the onset of the peeling of the film (maximum force = peeling force F_p). The process is then continued at a constant speed until peeling is complete. At the end of the test, it is necessary to check the state of both the band that has peeled off and that of the substrate, in order to ascertain that no lengthening or plastic deformation has occurred, and that all the energy expended during the peeling of the band has been used to break the bonds at the interface between the band and the substrate.

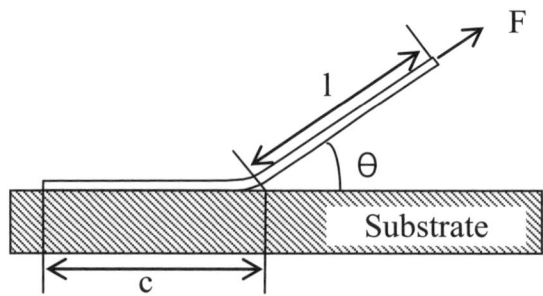

Figure 3.58. *Schematic representation of the peeling test principle*

In the case of an inextensible film of width b subjected to a peeling force F, the strain energy release rate can be written:

$$G = \left(\frac{F_p}{b}\right)(1 - \cos\theta) \qquad [3.54]$$

Peeling tests only yield reliable results in the case of low adherence. In fact, when adherence energy is high, the experimental results can be strongly affected by a significant extension of the band or by large dissipation of energy through plastic or viscoleastic deformation.

When the system tested does not behave strictly elastically, the peeling force will depend on the geometric and mechanical properties of the band and substrate [COG 00].

3.5.2.1.2. The blister test

The test consists of creating an opening in the substrate through the selective removal of material without altering the initial thickness of the coating. A pressure P is then applied to the film using a fluid (generally a gas), causing the film to deform, as shown in Figure 3.59. It gradually expands and blisters to a critical value δ_c corresponding to a pressure P_c before starting to peel off from the substrate. The strain energy release rate is given by the expression:

$$G = CP_c\delta_c \qquad [3.55]$$

with values for C generally ranging from 0.5 to 0.65 [BEN 04, COG 00]. P_c is expressed in MPa, δ in μm and G in J m^{-2}.

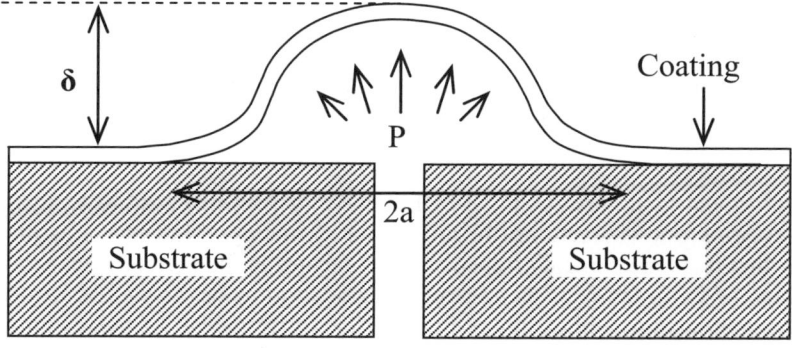

Figure 3.59. *Schematic representation of the blister test principle*

3.5.2.1.3. The scratch test

The *scratch test* [BUL 90, STE 85] applies a sharp indenter perpendicular to the surface to be treated with an applied load (F) that increases linearly with time. As the indenter penetrates the surface of the material, we subject the material to a sliding movement with a speed of a few millimeters per minute.

Scratch tests are usually carried out using a Rockwell indenter (a conical indenter with an apex angle of 120° and a radius of curvature of 200 μm), with a load ranging between 1 and 200 N, a loading rate of 10 N min^{-1} and a sliding speed of 10 mm min^{-1}.

Figure 3.60 shows the principle of this experimental set-up. The damage commonly observed during the sample motion falls into three main categories:

– semi-circular Hertzian cracks (at the beginning of the test: low load);

– cohesive scaling due to internal rupture within the film (medium load); and

– adhesive scaling due to interfacial rupture at the film-substrate interface (at the end of the test: high load).

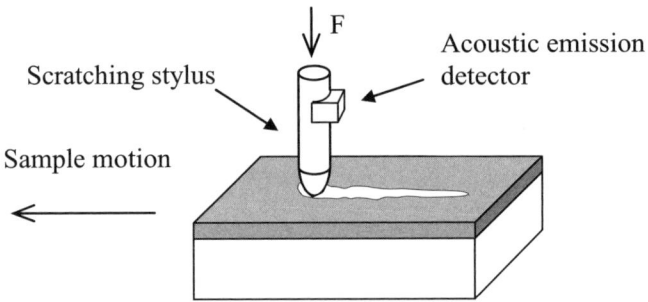

Figure 3.60. *Scratch test principle (adapted from [CSM 08])*

During the test, we generally measure the tangential force T and the acoustic emission (AE) generated during the cracking or scaling of the surface (see Figure 3.61).

The critical load corresponding to the onset of adhesive scaling is denoted Lc and can be used to characterize the adherence of coatings. However, it is important to note that there are some discrepancies and debates in the literature concerning the definition of Lc in terms of its particular degradation mode, as well as its detection and characterization. Some authors find that measurement of the tangential force is satisfactory, while others use primarily acoustic emission. Both methods involve observing the scratches and using optical or scanning electron microscopy in order check the state and mode of coating degradation.

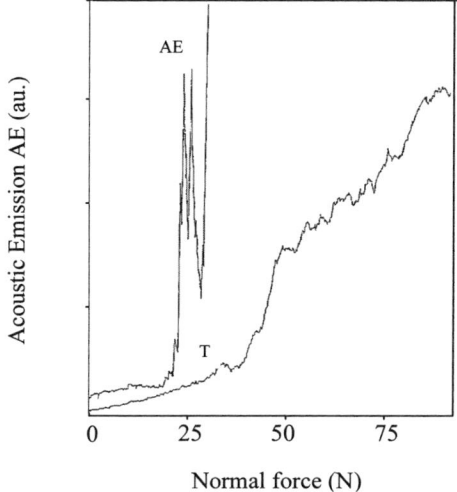

Figure 3.61. *Acoustic emission (AE) and tangential force measured during the scratch test*

Careful analysis of scratches has enabled five modes of coating degradation to be identified (see Figure 3.62) [BUR 87b]. The triggering of a particular degradation mode depends on many factors that can be divided into two categories: intrinsic factors that depend on the particular experimental conditions and extrinsic factors that depend on the nature and properties of the coating and substrate (see Table 3.8 below) [BUL 06].

For example, it is known that when the radius of curvature of the indenter, the thickness of the coating or the hardness of the substrate increase, it induces Lc to increase (see Figure 3.63). On the other hand, Lc decreases when the level of residual stresses present in the coating increases [BUL 90, BUL 01a].

Intrinsic	Extrinsic
Loading rate [BUL 89, STE 87, VAL 86]	Properties of the substrate (hardness, elastic modulus/elasticity) [STE 87]
Sliding speed [BUL 89, STE 87, VAL 86]	Properties of the coating (thickness, hardness, elastic modulus/elasticity, residual stresses) [BUL 89, STE 87]
Radius of curvature of the indenter [BUL 89, VAL 86]	Friction coefficient [THO 98, VAL 86]
Indenter wear [BUL 89]	Surface roughness [STE 87]
Stiffness / design of the measuring device	

Table 3.8. *Intrinsic and extrinsic factors impacting on the scratch test [BUL 06]*

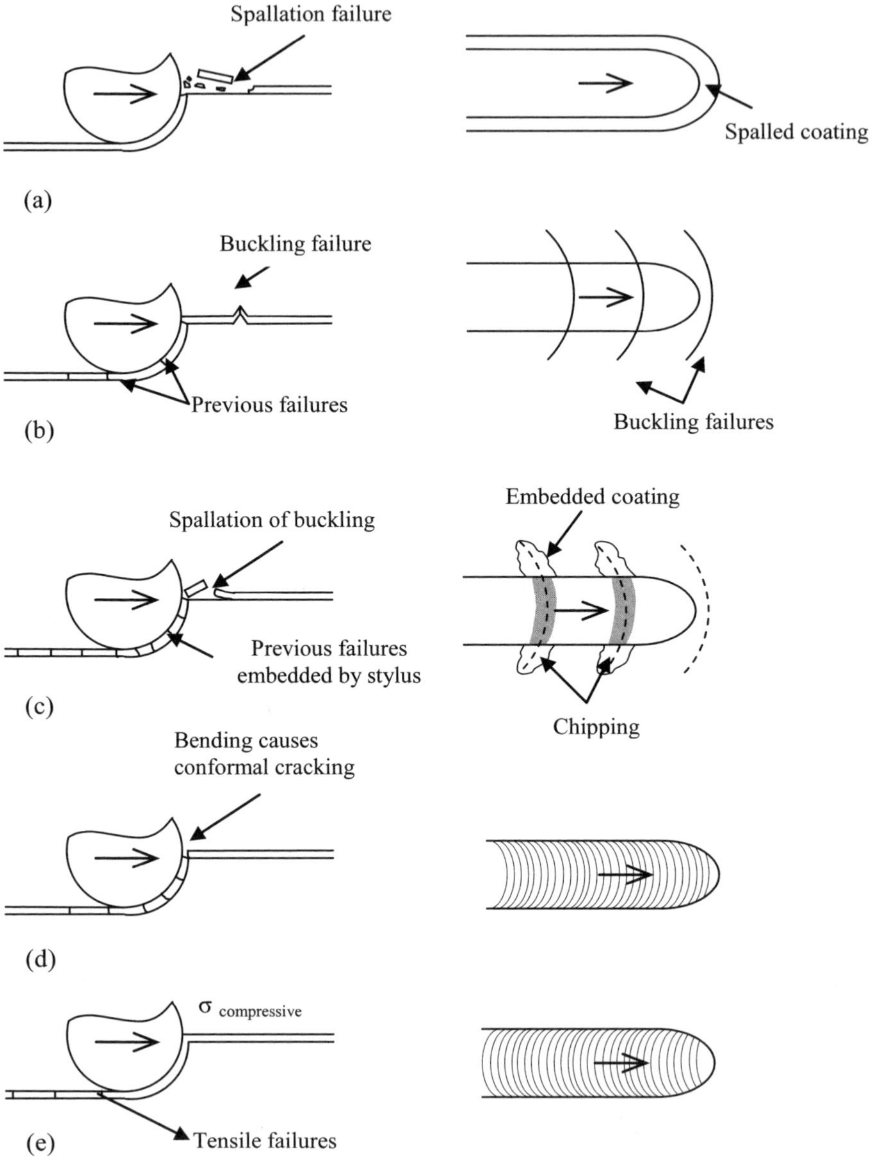

Figure 3.62. *Schematic representation of coating failure modes in scratch tests Plan and profile views of a) spalling failure; b) buckling failure; c) chipping failure; d) conformal cracking; e) tensile cracking [BUR 87b]*

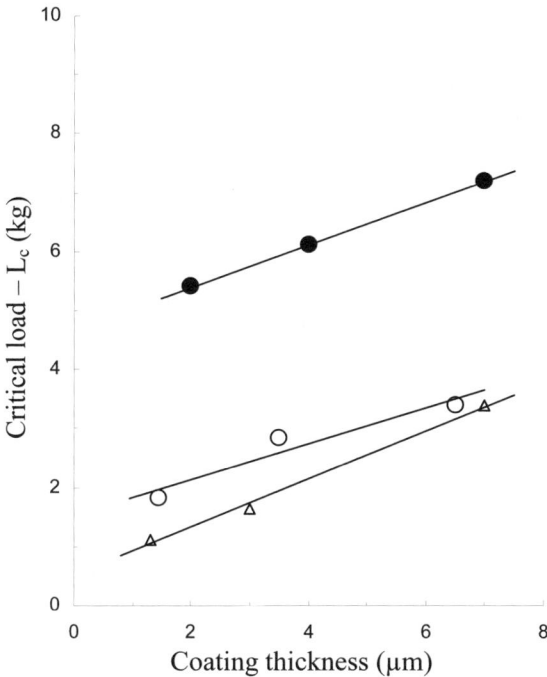

Figure 3.63. *Variation of Lc relative to the coating thickness of sputter-ion TiN deposited onto engineering steels: (△) stainless steel; (○) carbon steel; (●) tool steel (M2) [BUR 87b]*

When considering the impact of roughness on adherence, it can be said that increasing the substrate roughness makes it possible to improve the mechanical binding of the coating, and hence to increase Lc. However, increasing the roughness can also lead to the appearance of many defects likely to cause the onset and propagation of interfacial cracks.

The influence of the roughness on Lc is not completely clear, and the results reported by many authors are often contradictory, notably due to the particular choice of the method used to determine Lc. However, when roughness becomes too large, most authors notice degradation in the adherence.

Figure 3.64 presents results from a work focused on the characterization of the adherence of TiN films deposited onto steel substrates. It can clearly be seen that Lc is constant or increases slightly between Ra = 0.01 and 0.05 μm, but that it decreases significantly when the values for Ra are higher. A similar result was reported in [STE 87] where the authors observed a significant decrease in adherence when values for Ra are greater than 0.1 μm.

In another study [WIK 01], finite element modeling was used to show that high surface roughness generates significant stresses at the interface between the coating and substrate, which can cause coating detachment. The predictions of the model were verified by the qualitative analysis of many ceramic coatings (such as TiN, TiC, CrN, diamond or TiB_2) deposited onto steel or cemented carbides.

In [TAK 97a] the authors carried out a study of TiN films deposited onto steel substrates of varying roughness (Ra = 0.35 µm, Ra = 0.15µm and Ra = 0.02µm) and subjected them to scratch tests under constant loads. The analysis of these scratches under a scanning electron microscope clearly showed that adherence decreases as a function of the increase in roughness. Figure 3.65 shows an example of adhesive scaling obtained in the case of TiN film deposited on a steel substrate and scratched under an applied load of 5 N.

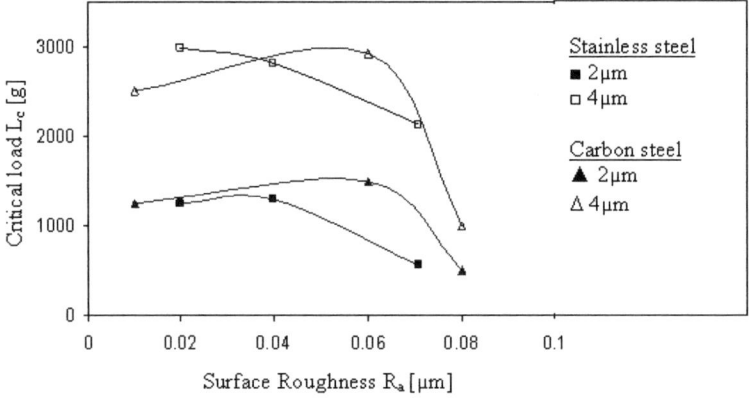

Figure 3.64. *Variation of Lc relative to the surface roughness of TiN films deposited onto stainless and carbon steels. L_c drops rapidly when R_a exceeds about 0.05 µm [BUL 90]*

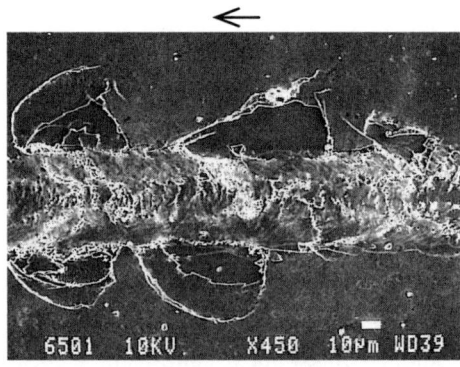

Figure 3.65. *Damage for scratch testing of a 3 µm thick TiN coating deposited by PVD onto a steel substrate. The applied load is 5 N and the arrow shows the sliding direction*

The different modes of degradation are classified as a function of the coating/substrate hardness ratio [BUL 97, BUL 06] (see Figure 3.66). For soft coatings (hardness < 5 GPa) deposited onto either hard or soft substrates, we see significant plastic deformation of the coating which leads to its rupture. However, for hard coatings (hardness > 5 GPa), two types of behavior may be observed:

– soft substrates (cohesive rupture within the coating); and

– hard substrates (lateral cracks and chipping as well as interfacial ruptures and adhesive scaling).

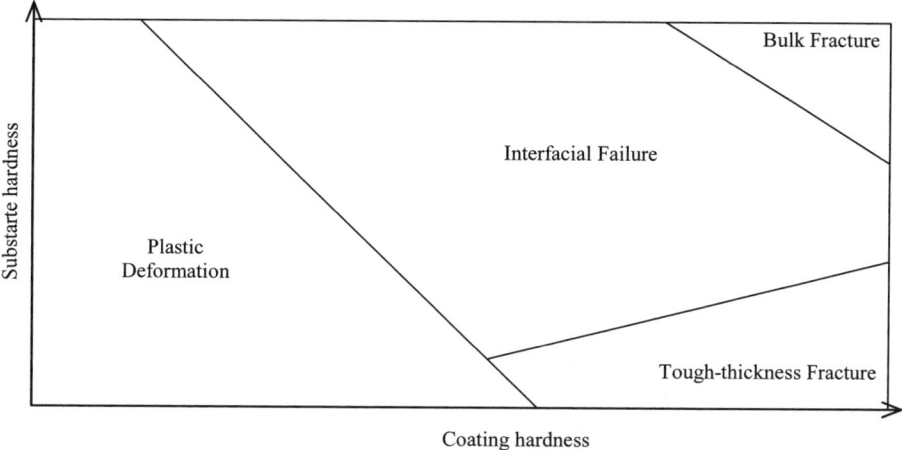

Figure 3.66. *Schematic representation of the dominating scratch test failure modes as a function of the coating and substrate hardness [BUL 06]*

The scratch test is characterized by the ease and speed with which it can be performed. However, the results it yields are to be considered with much caution. Many modes of material degradation can occur, with each requiring the definition of several corresponding critical forces.

Generally speaking, scratch tests are better suited to the characterization of adherence in hard coatings deposited on soft substrates. In these conditions, the scaling and rupturing of the film occur at the interface between the film and the substrate and the critical load Lc characterizes the adherence of the coating [BUL 06].

3.5.2.1.4. Interfacial indentation

This test requires the use of a pyramidal Vickers indenter to perform an indentation at the interface between the coating and the substrate (see Figure 3.67) [CHI 96]. Films of varying thickness are tested under increasing loads and the radial crack length is measured. If a logarithmic scale is used to plot the length of the crack as a function of the applied load, the points lie on a straight line. When several lines corresponding to the different films are plotted, the intersection of these lines corresponds to the critical load (P_c) of the crack onset, irrespective of the coating thickness (Figure 3.68). If a_c is the critical length of the half-diagonal of the indentation below which cracking does not occur, the interfacial toughness can be expressed using the relation:

$$K_c = 0.015 \frac{P_c}{a_c^{3/2}} \left(\frac{E}{H}\right)^{1/2} \qquad [3.56]$$

where

$$\left(\frac{E}{H}\right)^{1/2} = \frac{\left(\frac{E}{H}\right)_s^{1/2}}{1+\left(\frac{H_s}{H_f}\right)^{1/2}} + \frac{\left(\frac{E}{H}\right)_f^{1/2}}{1+\left(\frac{H_f}{H_s}\right)^{1/2}} \qquad [3.57]$$

where E_s, H_s, E_f, and H_f are the Young's moduli and hardness values for the substrate and the film, respectively.

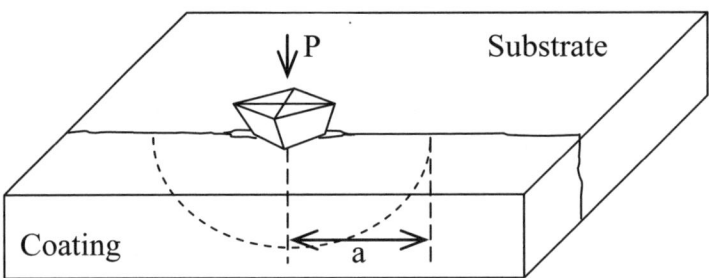

Figure 3.67. *Schematic representation of the interfacial indentation test*

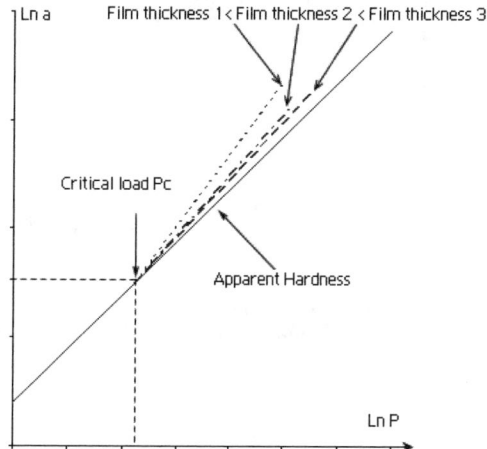

Figure 3.68. *Variation of Ln a relative to Ln P: a is the length of the half diagonal of the Vickers indentation and P is the applied load during the interfacial indentation test [CHI 96]*

3.5.3. Residual stresses in coatings

3.5.3.1. *Origin of internal stresses*

When a coating is produced at high temperatures (e.g. the thermal oxidation of silicon or PVD or CVD deposition of coatings), residual thermal stresses can appear in the deposited film as the materials return to ambient temperature. The stresses arise from the difference in the coating and substrate expansion coefficients, and the film–substrate system can deform if the substrate is sufficiently thin and ranges from a few tens to a few hundreds of microns.

For the case of a thick substrate, it is possible to reduce its thickness using either mechanical machining or chemical or electrochemical dissolution (Figure 3.69). When the substrate becomes sufficiently thin, the residual stresses cause the combined deformation of the film–substrate pair. The shape of the deformation – convex or concave – indicates tension or compression stresses, respectively.

The two main techniques used to determine the nature of internal stresses are X-ray diffraction and the Stoney method, based on measurements of the radius of curvature of the coating/substrate composite.

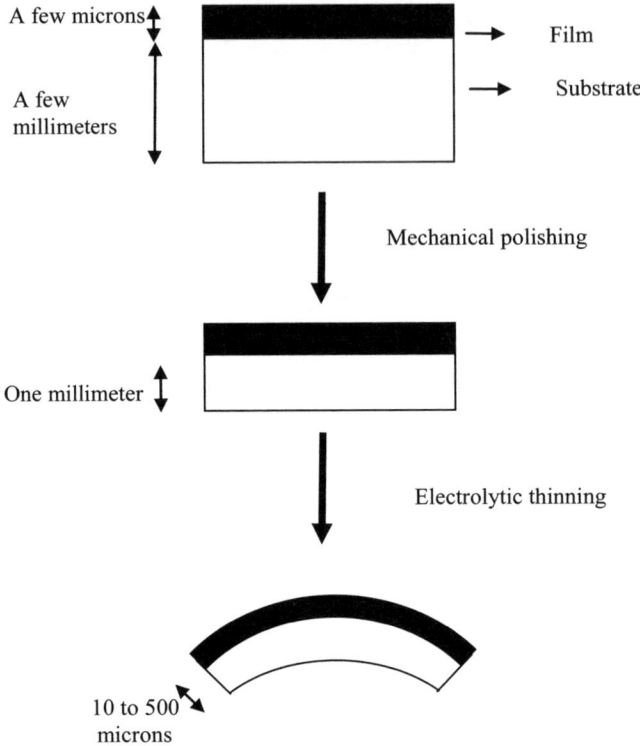

Figure 3.69. *Characterization of residual stresses within a coating after thinning of the substrate*

3.5.3.2. *Determining residual stresses using X-ray diffraction*

This technique, which is based on the measurement of the displacement of diffraction peaks, has been introduced in Chapter 1 (section 1.2.4.5.1).

3.5.3.3. *Determining internal stresses by radius of curvature measurements (Stoney's method)*

When the form of the coating-substrate system develops curvature under the effect of internal stresses, the mean stress σ in the film is calculated using [STO 09]:

$$\sigma = \frac{h_s^2 E_s}{6\rho h_f (1-v_s)} \qquad [3.58]$$

where h_s, υ_s and E_s represent the thickness, Poisson's coefficient and the Young's modulus of the substrate, respectively. The parameter h_f is the film thickness and ρ is the radius of curvature.

Note that this equation is only correct when the thickness ratio (h_f/h_s) is about 5–10%. For thick coatings, Stoney's equation must be modified as proposed by Roll [ROL 76].

Figure 3.70 shows the shape of three steel samples coated with a 3 µm thick film of TiN, TiCN or DLC (diamond-like carbon) deposited onto a thinned steel substrate (prepared as shown in Figure 3.71) using the PVD technique. A dozen points were measured using a mechanical profilometer and it was possible to determine the quadratic equation $y(x) = ax^2 + bx + c$ to best fit the measured deformation. The radius of curvature is simply given by the relation: $\rho = 1/y''(x) = 1/2a$ and the mean stress σ is calculated using equation [3.58] [HOU 98].

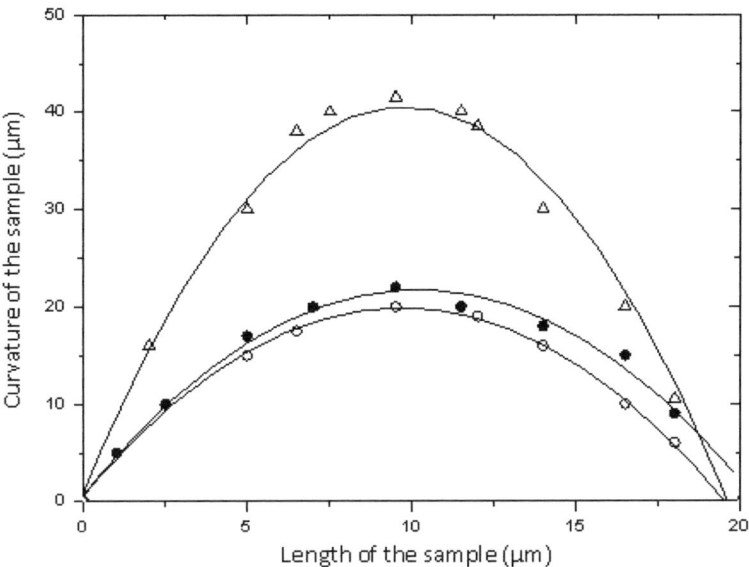

Figure 3.70. *Shape of coated steel substrates after thinning (--△--) TiCN, (--●--) TiN, (--○--) DLC [HOU 98]*

For the three coatings TiN, TiCN and DLC, the values generally published for internal stresses are generally the compression stresses, but they can vary greatly as a function of the experimental conditions when producing the film. There are many factors affecting internal stresses, such as the temperature of the substrate, the bias voltage applied to the substrate during deposition, the distance between the target material and substrate, the speed of deposition, the pressure and nature of the gases used or the mechanical and thermal properties of the substrate. Table 3.9 lists some values published by different authors.

Material	Internal stresses (MPa)	Publication Reference
TiN	− 2900	[RAM 90]
	−1500	[KIN 88]
	−2300	[HOU 98]
	[−2100, −50000]	[BUR 88]
	[−200, −10000]	[BUL 98]
	[−500, −5000]	[BUR 87c]
	−4000	[KIN 01]
	−3600	[HOU 98]
TiCN	−50000	[EIN 95]
	−3600	[HOU 98]
DLC	−1500	[WEI 82]
	−2000	[BAN 03]
	−2200	[HOU 98]

Table 3.9. *Measured internal stresses for several ceramic coatings*

Bibliography

[ABD 06] A. Abdel Aal, M. Khaled, K. Ibrahim, Z. Abdel Hamid, "Enhancement of wear resistance of ductile cast iron by Ni-SiC composite coating", *Wear*, vol. 260(9–10), 1070–1075, 2006

[ADD 06] H. Addach, M. Rezrazi, P. Berçot, J. Gavoille, M. Wery, H.F. Ayedi, J. Takadoum, "Elaboration et caractérisation mécanique et morphologique de revêtements électrolytiques de chrome", *Matériaux et techniques*, vol. 94(1), p. 47–52, 2006

[AGI 90] B. Agius, M. Froment *et al.*, *Surfaces interfaces et films minces*, Dunod, Paris, 1990

[AKI 04] T. Akihiro, N. Takayuki, S. Masahiro, M. Katsuhiro, "Characteristics of friction surfaces with DLC films in low and high humidity air", *Wear*, vol. 257(3–4), 297–303, 2004

[AMI 04] S. Amirthapandian, B.K. Panigrahi, S. Rajagopalans, A. Gupta, K.G.M. Nair, A.K. Tyagi, A. Narayanasamy, "Evidence for complete ion-beam mixing in thermally immiscible Fe/Ag multilayers from conversion electron Mössbauer spectroscopy", *Physical Review B. Condensed Matter and Materials Physics*, vol. 69(16), 165–411, 2004

[ANS 81] G.R. Anstis, P. Chantikul, B.R. Lawn, D.B. Marshall, "A critical evaluation of indentation techniques for measuring fracture toughness: I. Direct crack measurements", *Journal of the American Ceramic Society*, vol. 64(9), 533–538, 1981

[ANT 88] M. Antler, T. Spalvins, "Lubrication with thin gold films", *Gold Bulletin*, vol. 21(2), 59–68, 1988

[AYE 92] A. Ayel, M. Ganier, *Le lubrifiant véritable matériau de construction*, Cahiers du Cetim, Senlis, 1992

[BAN 85] H. Bangert, A. Wagendristel, "Ultra-load hardness tester for use in scanning electron microscope", *Review of Scientific Instruments*, vol 66, no. 8, p. 1568-1572, 1985

[BAN 03] M. Ban, T. Hasegawa, "Internal stress reduction by incorporation of silicon in diamond-like carbon film", *Surface and Coatings Technology*, vol. 162, p. 1-5, 2003

[BAR http] M. Barquins, "Adhésion et collage", http://es.ra.free.fr/art0022.php3

[BAR 01] S. Barril, S. Mischler, D. Landolt, "Triboelectrochemical investigation of the friction and wear behaviour of TiN coatings in a neutral solution", *Tribology International*, vol. 3(9), 599–608, 2001

[BEE 05] D. Beegan, M.T. Laugier, "Application of composite hardness models to copper thin hardness measurements", *Surface and Coatings Technology*, vol. 199, 32–37, 2005

[BEN 04] S. Benayoun, J.J. Hantzpergue, "Les tests d'adhérence appliqués aux revêtements minces: une synthèse bibliographique", *Matériaux et techniques*, (10–12), 23–31, 2004

[BERA 83] G. Béranger, H. Mazille, "Approche scientifique des surfaces. Caractérisation et propriétés", *Techniques de l'ingénieur*, M1425, 1–20, 1983

[BERA 94] G. Béranger, G. Henry, G. Sanz (ed.), *Le livre des aciers*, Tec & Doc, Paris 1994

[BERC 03] P. Berçot, "Dépôts composites par électrolyse – Modélisation", *Techniques de l'Ingénieur, Traité des matériaux métalliques M 1 622*, 2003

[BERTH 88] Y. Berthier, Mécanismes et tribologie, Thesis, INSA de Lyon, no. 88, INSAL 005, 1988

[BERTH 92] Y. Berthier, L. Vincent, M. Godet, "Velocity accommodation sites and modes in tribology", *European Journal of Mechanics–A/Solids*, vol. 11, 35–47, 1992

[BERTH 98] Y. Berthier, P. Kapsa, L. Vincent, in *Matériaux et contacts*, G. Zambelli, L. Vincent (ed.), Presses polytechniques et universitaires romandes, Lausanne, p. 2–12, 1998

[BERTH 05] Y. Berthier, "Third body reality, consequence and use of the third body to solve a friction and wear problem", in G. Stachowiack, *Wear, Materials, Mechanisms and Practice*, Wiley Intersciences, New York, p. 291–316, 2005

[BERTR 00] G. Bertrand, H. Mahdjoub, C. Meunier, "A study of the corrosion behaviour and protective quality of sputtered chromium nitride coatings", *Surface and Coatings Technology*, vol. 126(2–3), 199–209, 2000

[BHA 88] A.K. Bhattacharya, W.D. Nix, "Finite element analysis of cone indentation", *International Journal of Solids and Structures*, vol. 27(8), 1047–1058, 1991.

[BHU 91] B. Bhushan, B.K. Gupta, *Handbook of Tribology*, McGraw-Hill, New York, 1991

[BHU 05] B. Bhushan, *Nanotribology and Nanomechanics*, Springer Verlag, Berlin/Eldelberg/New York, 2005

[BIE 00] M. Bielmann, U. Mahajan, R.K. Singh, P. Agarwal, S. Mischler, E. Rosset, D. Landolt, *Materials Research Society Symposium*, Materials Research Society, vol. 566, 97–101, 2000

[BIN 86] G. Binnig, C.F. Quate, C. Gerber, "Atomic force microscopy", *Physical Review Letters.*, vol. 56, 930–933, 1986

[BIN 87] G. Binnig, C. Gerber, E. Stoll, T.R. Albrecht, C.F. Quate, "Atomic resolution with force microscope", *Europhysics Letters,* vol. 3, 1281–1286, 1987

[BLAN 01] E. Blando, E.K. Tentardini, R. Hubler, "Microhardness characterization of multilayers modified by ion beam mixing", *Nuclear Instruments and Methods in Physics Research*, Section B, vol. 175–177, 620–625, 2001

[BLAU 95] P.J. Blau, *Friction Science and Technology,* Marcel Dekker, New York, p. 7, 1995

[BOI 87] C. Boiziau, "Science des surfaces: recherche fondamentale et technologique", *Le vide, les couche minces*, (237) 259–261, May–July 1987

[BOUR 00] C. Bourdon, F. Leonidas, J. Takadoum, C. Roques-Carmes, "Etude du comportement de quelques matériaux soumis à la tribocorrosion ou à la corrosionérosion", *Tribologie et corrosion*, SIRPE, p. 80–84, 2000

[BOUS 05] N. Boussaa, A. Guittoum, S. Tobbeche, "Formation of Ni_2Si silicide in Ni/Si bilayers by ion beam mixing", *Vacuum*, vol. 77 (2), 125–130, 2005

[BOWD 50] F.P. Bowden, D. Tabor, *Friction and Lubrication of Solids*, Part I, Clarendon Press, Oxford, 1950

[BOWD 64] F.P. Bowden, D. Tabor, *Friction and Lubrication of Solids*, Part II, Clarendon Press, Oxford, 1964

[BOWE 86] K. Bowen, *Pour la science*, numéro spécial, no. 110, p. 126, 1986

[BRA 91] C. Brault, "L'anodisation des métaux légers ou semi-conducteurs", *Traitements de surface et composants mécaniques*, p. 213–241, Cetim, Senlis, 1991

[BRISCOEB 81] B.J. Briscoe, "Wear of polymers: an essay on fundamental aspects", *Tribology International*, vol. 14, 231–243, 1981

[BRISCOEB 92] B.J. Briscoe, T.A. Stolarski, "Friction", in W.A. Glaser (ed.), *Characterization of Tribological Materials*, Butterworth Publishers, London, p. 30–64, 1992

[BRISCOEE 98] E. Briscoe, A. Chateauminois, "Polymères" in G. Zambelli, L. Vincent (ed.), *Matériaux et contacts*, PPUR, Lausanne, p. 197–208, 1998

[BRU 03] C. Brun, M. Fromm, F. Berger, P. Delobelle, J. Takadoum, E. Berche, A. Chambaudet, F. Jaffiol, "Modifications of polypropylene surface properties by He+ ion implantation", *Journal of Polymer Science, Polymer Physics*, Part B, vol. 41(11), p. 1 183–1 191, 2003

[BUC 65] H. Buckle, *La machine outil française*, vol 206, p.125, 1965

[BUC 81] D.H. Buckley, *Surface Effects in Adhesion, Friction, Wear and Lubrication*, Elsevier, Amsterdam, 1981

[BUC 94] D.H. Buckley, K. Miyoshi, "Friction and wear of ceramics", *Wear*, vol. 100, 333–353, 1994

[BUL 89] S.J. Bull, D.S. Rickerby, A. Matthews, A.R. Pace, A. Leyland, "Scratch adhesion testing of hard, wear–resistant coatings", in Brozeit E, Munz W.D., Oechsner H, Riek K-T, Wolf GK, (editors), *Plasma Surface Engineering*, vol. 2, 1227–12235, 1989

[BUL 90] S.J. Bull, D.S. Rickerby, "New developments in the modelling of the hardness and scratch adhesion of thin films", *Surface and Coatings Technology*, vol. 42, p. 149–164, 1990

[BUL 95] S.J. Bull, "Tribology of carbon coatings: DLC, diamond and beyond", *Diamond Related Materials*, vol. 4(5–6), 827–836, 1995

[BUL 97] S.J. Bull, "Failure mode maps in the thin film scratch adhesion test", *Tribology International*, vol. 30, 491–498, 1997

[BUL 98] S.J. Bull, D.S. Rickerby, A. Matthews, A. Leyland, A.R. Pace, J. Perry, *Surface and Coatings Technology*, vol. 36, p. 503, 1998

[BUL 01a] S.J. Bull, "Modelling the hardness response of bulk materials, single and multilayer coatings", *Thin Solid film*, vol. 398–399, 291–298, 2001

[BUL 01b] S.J. Bull, D.S. Rickerby, "Characterization of hard coatings", in Bunshah R. F (ed.), *Handbook of Hard Coatings*, Noyes Publications, William Andrew Publishing, LLC, NY, USA, p. 181–228, 2001

[BUL 06] S.J. Bull, E.G. Berasetegui, "An overview of the potential of quantitative coating adhesion measurement by scratch testing", *Tribology International*, vol. 39(2), 99–114, 2006

[BUR 87a] P.J. Burnett, D.S. Rickerby, "The mechanical properties of wear–resistance coatings, I modelling of hardness behaviour", *Thin Solid Films*, vol. 148, 41–50, 1987

[BUR 87b] P.J. Burnett, D.S. Rickerby, "The relationship between hardness and scratch adhesion", *Thin Solid Films*, vol. 154, 403–416, 1987

[BUR 87c] P.J. Burnett, D.S. Rickerby, "The relationship between hardness and scratch adhesion", *Thin Solid Films*, vol. 154, 403–416, 1987

[BUR 88] P.J. Burnett, D.S. Rickerby, "The scratch adhesion test: an elastic–plastic indentation analysis", *Thin Solid Films*, vol. 157, 233–254, 1988

[BUZ 04] C. Buzea, K. Robbie, "Nano-sculptured thin film thickness variation with incidence angle", *Journal of Optoelectronics and Advanced Materials*, vol. 6, 1263–1268, 2004

[CAP 99] B. Cappela, G. Dietler, "Force-distance curves by atomic force microscopy", *Surface Science Reports*, vol. 34, 1–104, 1999

[CARRE 83] A. Carre, J. Schultz, "Polymer-aluminum adhesion. I. The surface energy of aluminum in relation to its surface treatment", *Journal of Adhesion*, vol. 15(2), 151–162, 1983

[CARREG 05] M. Carrega, *Aide-mémoire matières plastiques*, Dunod, Paris 2005

[CART 00] M. Cartier (ed.), *Guide d'emploi des traitements de surfaces appliqués aux problèmes de frottement*, Tec et Doc, Paris, 2000

[CAS 91] L. Castex, J.F. Flavenot, Y. Legurnic, "Le grenaillage de précontrainte – quels contrôles", *Traitements de surface et composants mécaniques*, p. 11–20, Cetim, Senlis, 1991

[CAT 05] J.M. Cattenot, *Communication orale, Deuxième Congrès de l'Institut des Traitements de Surface de Franche-Comté*, STIF2C, Besançon, 26–27 October 2005

[CEL 06] E. Celik, O. Culha, B. Uyulgan, N.F. Ak Azem, I. Ozdemir, A. Turk, "Assessment of microstructural and mechanical properties of HVOF sprayed WC-based cermet coatings for a roller cylinder", *Surface and Coatings Technology*, vol. 200 (14–15), 4320–4328, 2006

[CEN 92] Centre D'information Du Cuivre, *Laitons et Alliages les propriétés du cuivre et de ses alliages*, Variances, 1992

[CHA 88] M.M. Chaudhri, M. Winter, "The load-bearing area of a hardness indentation", *Journal of Physics D: Applied Physics*, vol. 21, 370–374, 1988

[CHAP 05] J.M. Chappe, Couches minces d'oxynitrures de titane par pulvérisation cathodique réactive, PhD Thesis, Franche-Comté University, 2005

[CHAS 95] E. Chassaing, M.P. Roumegs, M.F. Trichet, "Electrodeposition of Ni-Mo alloys with pulse reverse potentials", *Journal of Applied Electrochemistry*, vol. 25(7), 667–670, 1995

[CHE 95] N.G. Chechenin, J. Bottiger, J.P. Krog, "Nanoindentation of amorphous aluminium oxide films I. The influence of the substrate on the plastic properties", *Thin Solid Films*, vol. 261, 219–227, 1995

[CHENF 00] F. Chen, G.M. Brown, M. Song, "Overview of three dimensional shape measurement using optical method", *Optical Engineering*, vol. 39(1), 10–22, 2000

[CHENY 91] Y.M. Chen, J.C. Pavy, B. Rigaut, *Matériaux et techniques*, Special Issue, p. 40, December 1991

[CHI 95] D. Chicot, J. Lesage, "Absolute hardness of films and coatings", *Thin Solid Films* vol. 245, 123–130, 1995

[CHI 96] D. Chicot, P. Démarécaux, J. Lesage, "Apparent interface toughness of substrate and coating couples from indentation tests", *Thin Solid Films*, vol. 283, 151–157, 1996

[CHO 80] S.K.R. Chowdhury, N.E.W. Hartley, H.M. Pollock, M.A. Wilkins, "Adhesion energies at a metal interface: the effects of surface treatments and ion implantation", *Journal of Physics D: Applied Physics*, vol. 13, 1761–1784, 1980

[CHU 97] W.K. Chung, H.K. Kwang, "Anti-oxidation properties of TiAlN film prepared by plasma–assisted chemical vapor deposition and roles of Al", *Thin Solid Films*, vol. 307(1–2), 113, 1997

[COD 99] C. Coddet, G. Barbezat, P. Fauchais, G. Montavon, "Projection thermique et revêtements épais", in S. Audisio, M. Caillet, A. Galerie, H. Mazille (ed.), *Revêtements et traitements de surface*, Presses polytechniques et universitaires romandes, Lausanne, p. 467–477, 1999

[COG 00] J. Cognard, *Sciences et technologies du collage,* Presses polytechniques et universitaires romandes, Lausanne, 2000

[CONSTANT 92] A. Constant, G. Henry, J.C. Charbonnier, *Principes de base des traitements thermiques thermomécaniques et thermochimiques des aciers*, PYC, Ivry sur Seine, 1992

[CONSTANTI 06] R. Constanti, J. Matthey, S. Ramseyer, P.A. Steinmann, *Matériaux et Techniques*, vol. 94 (1), 11–21, 2006

[CSM 08] http://www.csm–instruments.com/frames/csemsom.html

[CUT 96] E.C. Cutiongco, D. Li, Y.W. Chung, C.S. Bhatia, "Tribological behavior of carbon nitride overcoats for magnetic film rigid disks", *Journal of Tribology*, vol. 118, 543–548, 1996

[DAR 97] E. Darque–Ceretti, "L'adhésion: les concepts et les causes: adhésionadhérence", *Revue de métallurgie*, vol. 94(5), 617–633, 1997

[DAR 03] E. Darque–Ceretti, E. Felder, *Adhésion et adhérence*, CNRS Editions, 2003

[DAV 88] D. David, R. Caplain (ed.), *Méthodes usuelles de caractérisation des surfaces*, Eyrolles, Paris, 1988

[DEGE 85] P.G. De Gennes, "Wetting: statics and dynamics", *Reviews of Modern Physics*, vol. 57(3), Part I, 827–863, July 1985

[DEGE 05] P.G. De Gennes, F. Brochard-Wyart, D. Quere, *Gouttes, bulles, perles et ondes*, Belin, Paris, 2005

[DEGR 98] A. De Graaf, G. Dinescu, J.L. Longueville, M.C.M. Van De Sanden, D.C. Schram, E.H.A. Dekempeneer, L.J. Van Ijzendoorn, "Amorphous hydrogenated carbon nitride films deposited via an expanding thermal plasma at high growth rates", *Thin Solid Films*, vol. 333, 29–34, 1998

[DER 75] B.V. Derjaguin, V.M. Muller, Y.J.P. Toporov, "Effect of contact deformations on the adhesion of particles", *Journal of Colloid Interface Science*, vol. 53(2), 314–326, 1975

[DES 87] M.C. Desjonqueres, D. Spanjaard, "Introduction à la physique des surfaces", *Le vide, les couches minces*, no. 237, 267–341, May–July 1987

[DIS 99] M. Diserens, J. Patscheider, F. Lévy, "Mechanical properties and oxidation resistance of nanocomposite TiN-SiNx physical-vapor-deposited thin films", *Surface and Coatings Technology*, vol. 120–121, 158–165, 1999

[DOW 76] D. Dowson, J.M. Challen, K. Holmes, J.R. Atkins, "The wear of non–metallic materials" in D. Dowson, M. Godet, C.M. Taylor (ed.), *Mechanical Engineering Publications*, 3rd Leeds-Lyon Symposium on Tribology, p. 99, 1976

[DUR 91] DURAND S., "Approche théorique et expérimentale de l'indentation des matériaux revêtus", Diplôme de DEA, université de Franche-Comté, 1991.

[EBE 89] J.P. Eberhart, *Analyse structurale et chimique des matériaux*, Dunod, Paris, 1989

[EIN 95] M. Einzenberg, K. Littau, S. Ghanayen, M. Liao, R. Moseley, A.K. Sinha, *Journal of Vacuum Science and Technology*, vol. A13, 590, 1995

[ERC 91] R.A. Erck, G.R. Frenske, A. Erdemir, "Uses of ion bombardment in thin-film deposition", forum given at *The Minerals & Materials Society Annual Meeting*, New Orleans, http://www.osti.gov/energycitations/servlets/purl/6458830–EY7kcj/6458830.PDF, 17–21 February 1991.

[EVA 79] D.C. Evans, J.K. Lancaster, "The wear of polymers, in wear", in D. Scott (ed.), *Treatise of Materials Science and Technology*, Academic Press, New York, vol. 13, p. 85, 1979

[FAY 87] S. Fayeulle, "The American Society of Mechanical Engineers", in K.C. Ludema (ed.), *International Conference on Wear of Materials*, New York, p. 13, 1987

[FIS 98] T.E. Fisher, M.P. Andersson, S. Jahanmir, R. Salher, "Friction and wear of though and brittle zirconia in nitrogen, air, water and hexadecane containing stearic acid", *Wear*, vol. 124(2), 133–148, 1998

[FLA 91] J.F. Flavenot, "Le grenaillage de précontrainte, influence des paramètres de l'opération sur les contraintes résiduelles introduites et sur la tenue à la fatigue", *Traitements de surface et composants mécaniques*, p. 21–47, Cetim, 1991

[FLE 97] P.D. Fleischauer, *New Directions in Tribology*, Mechanical Engineering Publications Ltd, p. 217, 1997

[FOW 67] F.M. Fowkes, "Surface chemistry", in R. L. Patrick (ed.), *Treatise of Adhesion and Adhesives*, vol. 11, Theory, Marcel Dekker, New York, p. 325–449, 1967

[FRE 04] C. Frétigny, "Les microscopies à champs de force", in M. Lehman, C. Dupras et P. Houdy (ed.), *Les nanosciences*, Belin, Paris, p. 94–119, 2004

[FRI 95] K. Friedrich, Z. Lu, A.M. Hager, "Recent advances in polymer composites' tribology", *Wear*, vol. 190(2), 139–144, 1995

[FRI 98] K. Friedrich, "Composite à matrice métallique", in G. Zambelli, L. Vincent (ed.), *Matériaux et contacts*, Presses polytechniques et universitaires romandes, p. 209–225, 1998

[GAL 02] A. Galerie, *Traitements de surfaces en phase vapeur*, Hermes, Paris, 2002

[GAN 04] W. Gang, L. Ning, Z. Derui, M. Kurachi, "Electrodeposited Co-Ni-Al2O3 composite coatings", *Surface and Coatings Technology*, vol. 176, 157–164, 2004

[GAR 01] I. Garcia, J. Fransaer, J.P. Celis, "Electrodeposition and sliding wear resistance of nickel composite coatings containing micron and submicron SiC particles", *Surface and Coatings Technology*, vol. 148 (2–3), 171, 2001

[GAS 95] Gasvik K.J., *Optical Metrology*, John Wiley & Sons, New York, 2nd edition, 1995

[GAV 02a] J. Gavoille, J. Takadoum, "Study of surface force dependence on pH by atomic force microscopy", *Journal of Colloid and Interface*, vol. 250, 104–107, 2002

[GAV 02b] J. Gavoille, Caractérisation des propriétés d'adhésion et de frottement de quelques matériaux: influence des conditions opératoires, PhD Thesis, Franche-Comté University, Besançon, 2002

[GAY 01] P.A. Gay, P. Berçot, J. Pagetti, *Surface and Coatings Technology*, vol. 140, p. 147, 2001

[GEO 94] J.M. Georges, A. Tonck, D. Mazuyer, "Interfacial friction of wetted monolayers", *Wear*, vol. 175(1–2), 59–62, 1994

[GEO 00] J.M. Georges, *Frottement, usure et lubrification*, éditions CNRS/Eyrolles, Paris, 2000

[GRE 66] J.A. Greenwood, J.B.P.Williamson, "Contact of nominally flat surfaces", *The Royal Society*, London, vol. A295, 300–319, 1966

[GRI 93] A. Grill, "Review of the tribology of diamond-like carbon", *Wear*, vol. 168(1–2), 143–153, 1993

[GROG 92] A.L. Grogan, V.H. Desai, F.C. Gray, S.L. Rice, "Apparatus for chemomechanical wear studies with biaxial load and surface charge control", *Wear*, vol. 152(2), 383–393, 1992

[GROS 01] A. Grosjean, M. Rezrazi, J. Takadoum, P. Berçot, "Hardness, friction and wear characteristics of nickel-SiC electroless composite deposits", *Surface and Coatings Technology*, vol. 137, 92–96, 2001

[GUI 05] X. Guijun, Z. Zhong, "Sliding wear of polyetherimide matrix composites: I. Influence of short carbon fibre reinforcement", *Wear*, vol. 258(5–6), 776–782, 2005

[HAE 02] H. Haefke, S. Ortmann, Y. Gerbig, A. Savan, *10th Nordic Symposium on Tribology*, NORDTRIB, Stockholm, Sweden, 9–12 June 2002

[HALI 02] D.N. Haliyo, Les forces d'adhésion et les effets dynamiques pour la micromanipulation, PhD Thesis, Pierre et Marie Curie University, Paris, 2002

[HALL 86] J. Halling, "The tribology of surface coatings, particularly ceramics", *Institution of Mechanical Engineers*, vol. 200, 31–40, 1986

[HAM 95] R. Hamzah, D.J. Stephenson, J.E. Strutt, "Erosion of material used in petroleum production", *Wear*, (186–187), 493–496, 1995

[HAN 87] Z. Hanmin, H. Guoren, Y. Guicheng, "Friction and wear of poly(phenylene sulphide) and its carbon fibre composites: I unlubricated", *Wear*, vol. 116, 59–68, 1987

[HOU 98] H. Houmid Bennani, "Contribution à l'étude de l'adhérence et des caractéristiques mécaniques et tribologiques de quelques couches minces dures", PhD thesis, Franche-Comté University, Besançon. Numéro d'ordre 632, 1998

[HUA 94] Z.P. Huang, Y. Sun, T. Bell, "Friction behaviour of TiN, CrN and (TiAl)N coatings", *Wear*, vol. 173, 13–20, 1994

[HUB 01] R. Hubler, "Ion beam mixing of Ti-TiN multilayers for tribological and corrosion protection", *Nuclear Instruments and Methods in Physics Research*, Section B, vol. 175–177, 630–636, 2001

[HUM 83] E. Hummer, J. Perry, "Adhesion and hardness of ion-plated TiC and TiN coatings", *Thin Solid Films*, vol. 101, p.243-251, 1983

[HUT 92] I.M. Hutchings, *Tribology: Friction and Wear of Engineering Materials*, Edward Arnold, London, 1992

[IOS 96] A. Iost, R. Bigot, "Indentation size effort: reality or artefact", *Journal of Materials Science*, vol 31, 3573–3577, 1996

[IOS 05] A. Iost, Y. Ruderman, M. Bigerelle M., "Dureté des revêtements: quel modèle choisir?", *Matériaux et techniques*, vol. 93, 201–211, 2005

[ISR 94] J.N. Israelachvili, Y.L. Chen, H. Yoshizawa, "Relation adhesion and friction forces", *Journal of Adhesion Science and Technology*, vol. 8 (11), 1 231–1 249, 1994

[ISR 99] J.N. Israelachvili, *Intermolecular and Surface Forces*, Academic Press, London, 1999

[JIA 93] X.X. Jiang, S.Z. Li, D.D. Tao, J.X. Yang, "Accelerative effect of wear on corrosion of high–alloy stainless steel", *Corrosion*, vol. 49 (10), 836–841, 1993

[JOH 71] K.L. Johnson, K. Kendall, A.D. Roberts, "Surface energy and the contact of elastic solids", *The Royal Society*, London, vol. A324, 301–313, 1971

[JOH 85] K.L. Johnson, *Contact Mechanics*, Cambridge University Press, Cambridge, 1985

[JON 84] B. Jonsson, S. Hogmark, "Hardness measurements of thin films", *Thin Solid Films*, vol 114, 257–269, 1984

[JOS 95] A. Joshi, H.S. Hu, "Oxidation behavior of titanium-aluminium nitrides", *Surface and Coatings Technology*, vol. 76–77, 499–507, 1995

[KAB 95] A. Kabuya, C. Haesen, *Matériaux et Techniques*, vol. 1–2, p. 31, 1995

[KHU 96] A. Khurshudov, K. Kato, D. Swada, "Tribological and mechanical properties of carbon nitride thin coating prepared by ion-beam-assisted deposition", *Tribology Letters*, vol. 2(1),13–21, 1996

[KIN 88] A. Kinbara, S. Baba, "Adhesion measurement of non-metallic thin films using a scratch method", *Thin Solid Films*, vol. 163, p. 67–73, 1988

[KIN 01] H. W. King, T. A. Caughlin, D. R. Nagy, "Residual stress, hardness and chemical stability of TiN coatings", *Journal of Advanced Materials*, vol. 33, no. 1, p. 63-68, 2001

[KON 97] K. Kondo, M. Yokoyama, K. Shinohara, "Morphology evolution of zinc–nickel binary alloys electrodeposited with pulse current", *Journal of Electrochemical Society*, vol. 142(7), 2256–2260, 1997

[KOR 98] A.M. Korsunsky, M.R. McGurk, S.J. Bull, T.F. Page, "On the hardness of coated systems", *Surface and Coatings Technology*, vol. 99, 171–183, 1998

[KRI 96] J. Krim, "Les frottements à l'échelle atomique", *Pour la Science*, no. 230, p. 54–60, décembre 1996

[KUS 98] Y. Kusano, J.E. Evetts, R.E. Somekh, I.M. Hutchings, "Properties of carbon nitride films deposited by magnetron sputtering", *Thin Solid Films*, vol. 332(1–2), 56–61, 1998

[LAN 90] J.K. Lancaster, "A review of the influence of environmental humidity and water on friction, lubrication and wear", *Tribology International*, vol. 23 (6), 371–389, 1990

[LAV 93] P. Laval, E. Felder, "Caractérisation de l'adhérence des revêtements par indentation normale: une revue bibliographique", *Matériaux et techniques*, 1–3, 93–105, 1993

[LAW 80] B.R. Lawn, A.G. Evans, D.B. Marshall, "Elastic plastic indentation damage in ceramics: The median radial crack system", *Journal of the American Ceramic Society*, vol. 63, 574–581, 1980

[LEB 85] D. Lebouvier, P. Gilormini, E. Felder, "A kinematic solution for plane–strain indentation of a bi–layer", *Journal of Physics; D: Applied Physique*, vol. 18, 199–210, 1985

[LEH 96] T. Le Huu, H. Zaidi, D. Paulmier, P. Voumard, "Transformation of sp3 to sp2 sites of diamond like carbon coatings during friction in vacuum and under water vapour environment", *Thin Solid Films*, vol. 290–291, 126–130, 1996

[LEV 94] R. Leveque, M. Entringer, "Les propriétés intrinsèques et fonctionnelles des surfaces" in G. Béranger, G. Henry and G. Sanz (ed.), *Le livre de l'acier*, Tec et Doc, Lavoisier, Paris, p. 584–608, 1994

[LI 92] S.Z. Li, Y. Shi, H. Peng, *Plasma Chem. Process.*, vol. 12, p. 287, 1992

[LIE 87] H.P. Lieulade, "Effet des contraintes résiduelles introduites par des traitements de surface sur le comportement à la fatigue", in A. Niku–Lari (ed.), *Traitements et revêtements des métaux*, p. 201–236, Hermès, Paris, 1987

[LIG 04] J.L. Ligier, *Avaries en lubrification*, Technip, Paris, p. 15, 2004

[LIM 87] S.C. Lim, M.F. Ashby, J.H. Brunton, *Acta metallurgica*, vol. 35, 1343–1348, 1987

[LIM 97] S.C. Lim, "Recent developments in wear-mechanism maps", in I.M. Hutchings (ed.), *New Directions in Tribology*, Mechanical Engineering Publications Ltd, London, p. 309–320, 1997

[LIN 03a] J. Lintymer, Etude de l'influence de la microstructure sur les propriétés mécaniques et électriques des couches de chrome en zigzag élaborées par pulvérisation cathodique, PhD Thesis, Franche-Comté University, Besançon, 2003

[LIN 03b] J. Lintymer, J. Gavoille, N. Martin, J. Takadoum, "Glancing angle deposition to modify microstructure and properties of sputter deposited chromium thin films", *Surface and Coatings Technology*, vol. 174–175 C, 316–323, 2003

[LIN 04] J. Lintymer, N. Martin, J.-M. Chappé, P. Delobelle, J. Takadoum, "Influence of zigzag microstructure on mechanical and electrical properties of chromium multilayered thin films", *Surface and Coatings Technology*, vol. 180–181, 26–32, 2004

[LIN 06] J. Lintymer, N. Martin, J.-M. Chappé, J. Takadoum, P. Delobelle, "Modeling of Young's modulus, hardness and stiffness of chromium zigzag multilayers sputter deposited", *Thin Solid Films*, vol. 503(1–2), 177–189, 2006

[LIUA 89] A.Y. Liu, M.L. Cohen, "Prediction of new low compressibility solids", *Science*, vol. 245, 841–842, 1989

[LIUA 90] A.Y. Liu, M.L. Cohen., "Structural properties and electronic structure of low compressibility materials: β-Si_3N_4 and hypothetical β-C_3N_4", *Physical Revue*, vol. B41, 10,727–10,734, 1990

[LIUC 01] C. Liu, Q. Bi, A. Matthews, "EIS comparison on corrosion performance of PVD TiN and CrN coated mild steel in 0.5 N NaCl aqueous solution", *Corrosion Science*, vol. 43(10), 1953–1961, 2001

[LIUY 96] Y. Liu, A. Erdemi, E.I. Meletis, "A study of the wear mechanism of diamondlike carbon films", *Surface and Coatings Technology*, vol. 82, 48–56, 1996

[LOS 99] B.B. Losiewicz, A. Stepien, D. Gierlotka, A. Budniok, "Composite layers in Ni–P system contining TiO_2 and PTFE", *Thin Solid Films*, vol. 349(1–2), 43–50, 1999

[LUZ 95] Z.P. Lu, K. Friedrich, "On sling friction and wear of PEEK and its composites", *Wear*, vol. 181–183, 624–631, 1995

[MAC 86] E. Macherauch, V. Hauk (ed.), *Residual Stresses in Science and Technology*, DGM, 1986

[MAE 88] G. Maeder, J.L. Lebrun, "Mesure de contraintes", in D. David and R. Caplain (ed.), *Méthodes usuelles de caractérisation des surfaces*, Eyrolles, Paris, p. 251–269, 1988

[MANA 04] C. Manasterski, *La pulvérisation cathodique industrielle*, Presses polytechniques et universitaires romandes, Lausanne, 2004

[MANG 01] T. Mang, W. Dresel (ed.), *Lubricants and lubrication*, Wiley-Vech, Weinheim, 2001

[MARTI 95] A. Marti, G. Hahner, N.D. Spencer, "Sensitivity of frictional forces to pH on a nanometer scale: a lateral force microscopy study", *Langmuir*, vol. 11, 4632–4635, 1995

[MARTIN 02] N. Martin, J. Lintymer, J. Gavoille, J. Takadoum, "Nitrogen pulsing to modify the properties of titanium nitride thin films sputter deposited", *Journal of Materials Science*, vol. 37(20), 4327–4332, 2002

[MAS 99] H.H. Masjuki, M.A. Maleque, A. Kubo, T. Nonaka, "Palm oil and mineral oil based lubricants: their tribological and emission performance", *Tribology International*, vol. 32(6), 305–314, 1999

[MAT 87] C.M. Mate, G.M. Mccleliand, R. Erlandsson, S. Chiang, "Atomic-scale friction of a tungsten tip on a graphite surface", *Physical Review Letters*, vol. 59, 1942–1945, 1987

[MAU 78] D. Maugis, M. Barquins, "Fracture mechanics and the adherence of viscoelastic bodies", *Journal of Physics D: Applied Physics*, vol. 11, 1989–2023, 1978

[MAU 84] D. Maugis, *Le vide, les couches minces*, vol. 220, p. 3–21, 1984

[MEN 00] P.F. Mentone, *AESF 5th International Pulse Plating Symposium*, Chicago, 29–30 June 2000

[MES 84] R. Messier, A.P. Giri, R.A. Roy, "Revised structure zone model for thin film physical structure", *Journal of Vacuum Science and Technology*, vol. A2, 500–503, 1984

[MIC 89] A. Michel, "Caractérisation et mesure des microgéométries de surface", *Techniques de l'ingénieur*, R1230, 1–20, 1989

[MIN 98] P. Minotti, A. Ferreira, *Les Micromachines*, Hermès, Paris, 1998

[MIS 93] S. Mischler, E. Rosset, G.W. Stachowiak, D. Landolt, "Effect of sulphuric acid concentration on the rate of tribocorrosion of iron", *Wear*, vol. 167(2), 101–108,1993

[MIS 99] S. Mischler, A. Spiegel, D. Landolt, "The role of passive oxide films on the degradation of steel in tribocorrosion systems", *Wear*, vol. 225–229, 1078–1087, 1999

[MIY 82] K. Miyoshi, D. Buckley, *ASLE Transactions*, vol. 27(1), 51, 1982

[MIY 90] K. Miyoshi, "Studies of mechanochemical interactions in the tribological behaviour of materials", *Surface and Coatings Technology*, vol. 43–44, 799–821, 1990

[MOON 91] H. Moon–Hee, I.P. Su, "Corrosive wear behaviour of 304–L stainless steel in 1 N H2SO4 solution Part 1. Effect of applied potential", *Wear*, vol. 147(1), 59–67, 1991

[MOOR 81] M.A. Moore, in D.A. Rigney (ed.), *Fundamentals of Friction and Wear of Materials*, ASM, p. 73, 1981

[MOR 04] C. Morant, P. Prieto, A. Forn, J.A. Picas, E. Elizalde, J.M. Sanz, "Hardness enhancement by CN/TiCN/TiN multilayers films", *Surface and Coatings Technology*, vol. 180–181, 512–518, 2004

[MYS 97] N.K. Myshkin, C.K. Kim, M.I. Petrokovets, *Introduction to Tribology*, p. 88, Cheong Moon Gak, Seoul, 1997

[NAV 93] B. Navinsek, P. Panjan, "Oxidation resistance of PVD Cr, Cr–N and Cr–N–O hard coatings", *Surface and Coatings Technology*, vol. 59(1–3), 244–248, 1993

[NEV 99] A. Neville, T. Hodgkiess, H. Xu, "An electrochemical and microstructural assessment of erosion–corrosion of cast iron" *Wear*, vol. 233–235, 523–534, 1999

[NGU 98] B. Nguyen, "Electrodéposition par courants pulsés", *Techniques de l'Ingénieur, Traité des matériaux*, M1627, 1998

[NIN 05] L. Ning, H. Chengliang, Y. Haidong, X. Yudong, S. Min, C. Sheng, X. Feng, "The milling performances of TiC–based cermet tools with TiN nanopowders addition against normalized medium carbon steel AISI1045", *Wear*, vol. 258(11–12), 1688–1695, 2005

[NIX 98] W.D. Nix, H. Gao, "Indentation size effects in crystalline materials: a law for strain gradient plasticity', *Journal of Mechanics and Physics of Solids*, vol, 411–425, 1998

[OLI 92] W.C. Olivier, G.M. Pharr, "An improved technique for determining hardness and elastic modulus using load and displacement sensing indentation experiments", *Journal of Materials Research*, vol. 7(6), 1564–1583, 1992

[OLI 04] W.C. Olivier, G.M. Pharr, "Measurement of hardness and elastic modulus by instrumented indentation: advances in understanding and refinements to methodology", *Journal of Materials Research*, vol. 19(1), 3–20, 2004

[OMA 86] M.K. Omar, A.G. Atkins, J.K. Lancaster, "The role of crack resistance parameters in polymer wear", *Journal of Physics D: Applied Physics*, vol. 19, 177–195, 1986

[OUD 73] J. Oudar, *La chimie des surfaces*, PUF, Paris, 1973

[OWE 69] D.K. Owens, R.C. Wendt, "Estimation of the surface free energy of polymers", *Journal of Applied Polymer Science*, vol. 13, 1741–1747, 1969

[PASH 84] M.D. Pashley, J.B. Pethica, D. Tabor, "Adhesion and micromechanical properties of metal surfaces", *Wear*, vol. 100, 7–13, 1984

[PAST 87] H. Pastor, *Matériaux et Techniques*, p. 319, July/août, 1987

[PAU 96] Y. Pauleau, *Les revêtements anti-frottement*, TechTendances, Paris, Innovation 128, 1996

[PAW 03] L. Pawlowski, *Dépôts physiques*, Presses polytechniques et universitaires romandes, Lausanne, 2003

[PEN 98] E. Peña Muñoz, P. Berçot, A. Grosjean, M. Rezrazi, J. Pagetti, "Electrolytic and electroless coatings of Ni-PTFE composites: study of some characteristics", *Surface and Coatings Technology*, vol. 107, 85–93, 1998

[PEP 76] S. Pepper, *Journal of Applied Physics*, vol. 47, no. 3, 1976

[PERR 87] B. Perraillon, "Les surfaces en métallurgie", *Le vide, les couches minces*, vol. 42 (238), 423–426, 1987

[PERS 98] B.N.J. Persson, *Sliding Friction*, Springer Verlag, New York, 1998

[PET 06] C. Petitjean, M. Grafoute, C. Rousselot, J.F. Pierson, O. Banakh, A. Cavaleiro, *Matériaux et Techniques*, vol. 94(1), 23, 2006

[PIE 99] O. Piétrement, J.L. Beaudoin, M. Troyon, "A new calibration method of the lateral contact stiffness and lateral force using modulated lateral force microscopy", *Physical Review B. Condensed Matter*, vol. 7(4), 213–220, 1999

[PIL 06] D. Pilloud, J.F. Pierson, J. Takadoum, "Structure and tribological properties of reactively sputtered Zr–Si–N films", *Thin Solid Films*, vol. 496(2), 445–449, 2006

[PIR 06] J. Pirso, M. Viljus, S. Letunovits, "Friction and dry sliding wear behaviour of cermets", *Wear*, vol. 260, 851–824, 2006

[PIV 87] J.C. Pivin, F. Pons, J. Takadoum, H.M. Pollock, G. Farges, "Study of the correlation between hardness and structure of nitrogen–implanted titanium surfaces", *Journal of Materials Science*, vol. 22, 1087–1096, 1987

[PIV 94] J.C. Pivin, "Hardening and embrittlement of polyimide by ion implantation", *Nuclear Instruments and Methods in Physics Research B*, vol. 84(4), 484–490, 1994

[POD 04] B. Podgornik, J. Vizintin, S. Jacobson, S. Hogmark, "Tribological behaviour of WC/C coatings operating under different lubrication regimes", *Surface and Coatings Technology*, vol. 177, 558–565, 2004

[POL 77] H.M. Pollock, P. Shufflebottom, J. Skinner, "Contact adhesion between solids in vacuum. I. Single-asperity experiments", *Journal of Physics D: Applied Physics*, vol. 10, 127–138, 1977

[PON 04] P. Ponthiaux, F. Wegner, D. Drees, J.P. Celis, "Electrochemical techniques for studying tribocorrosion processes", *Wear*, 459–468, 2004

[POS 05] K. Poser, K.H. Zum Gahr, J. Schneider, "Development of Al2O3 based ceramics for dry friction systems", *Wear*, vol. 259(1–6), 529–538, 2005

[PUC 02] E.S. Puchi-Cabrera, "A new model for the computation of the composite hardness of coated systems", *Surface and Coatings Technology*, vol. 160, 177–186, 2002

[QIU 02] C. Qiuying, M. Yonggang, W. Shizhu, "Influence of interfacial potential on the tribological behavior of brass/silicon dioxide rubbing couple", *Applied Surface Science*, vol. 202, 120–125, 2002

[RAM 90] H.W. Ramsey, H.W. Chandler, T.F. Page, "The determination of residual stresses in thin coatings by a sample thinning method", *Surface and Coatings Technology*, vol. 43–44, 223–233, 1990

[RAU 00] J.Y. Rauch, N. Martin, C. Rousselot, C. Jacquot, J. Takadoum, "Characterization of (Ti1-xAlx)N films prepared by radio frequency reactive magnetron sputtering", *Journal of the European Ceramic Society*, vol. 20, 795–799, 2000

[REZ 05] M. Rezrazi, M.L. Doche, P. Berçot, J.Y. Hihn, "Au-PTFE composite coatings elaborated under ultrasonic stirring", *Surface and Coatings Technology*, vol. 192, 124–130, 2005

[RIC 94] A. Richard, A.M. Durand, *Le vide, les couches minces, les couches dures*, In Fine, Paris, 1994.

[RIV 95] J.P. Rivière, M. Zaytouni, M.F. Denanot, J. Allain, *Material Science Engineering B*, vol. 29(1–3), 105, 1995

[RIV 99] J.P. Riviere, "Traitement et revêtements de surface par faisceaux d'ions", in S. Audisio, M. Caillet, A. Galerie, H. Mazille (ed.), *Revêtements et traitements de surface*, p. 520–538, Presses polytechniques et universitaires romandes, Lausanne, 1999

[RIV 00] J.P. Rivière, *Journal de Physique IV*, vol. 10, p. 53, 2000

[ROC 91] A. Roche, "Les theories de l'adhésion et mesure de l'adhérence", *Le Vide, les couches minces*, no. 257, May, June, July 1991

[ROL 76] K. Roll, "Analysis of stress and strain distribution in thin films and substrates", *Journal of Applied Physics*, vol. 47(7), 3224–3229, 1976

[ROL 00] Y. Rollot, S. Regnier, "Micromanipulation par adhésion. Modélisation dynamique et expérimentations", *Nano et micro technologies*, vol. 1(2), 213–241, 2000

[RUD 03] L.R. Rudnick (ed.), *Lubricant Additives*, Marcel Dekker, New York, 2003

[SAR 79] P.M. Sargent, PhD Thesis, University of Cambridge, UK, 1979

[SATOM 00] M Sato, M. Brett, *Vacuum Solutions*, vol. 14, p. 26–31, 2000

[SATOT 94] T. Sato, Y. Tada, M. Ozaki, T. Hoke, T. Besshi, "A crossed-cylinders testing for evaluation of wear and tribological properties of coated tools", *Wear*, vol. 178(1–2), 95–100, 1994

[SCH 97a] U.D. Schwarz, O. Zworner, P. Koster, R. Wiesendanger, "Quantitative analysis of the frictional properties of solid materials at low loads, I. carbon compounds", *Physical Review B*, vol. 56(11), 6987–6996, 1997

[SCH 97b] U.D. Schwarz, O. Zworner, P. Koster, R. Wiesendanger, "Quantitative analysis of the frictional properties of solid materials at low loads., II. mica and germanium sulfide", *Physical Review B*, vol. 56(11), 6997–7000, 1997

[SET 99] M.W. Seto, K. Robbie, D. Vick, M.J. Brett, L. Kuhn, "Mechanical response of thin films with helical microstrucures", *J. Vac. Sci. Technol.* B17(5), 2172–2177, 1999

[SET 01] M.W. Seto, B. Dick, M.J. Brett, "Mechanical response of microsprings and microcantilevers", *J. Micromechanics and Microengineering*, 11, 582–588, 2001

[SHI 05] P.H. Shipway, L. Howell, "Microscale abrasion-corrosion behaviour of WC-Co hard metals and HVOF sprayed coatings", *Wear*, vol. 258, 303–312, 2005

[SHR 01] K. Shrestha Nabeen, K. Sakurada, M. Masuko, T. Saji, "Composite coatings of nickel and ceramic particles prepared in two steps", *Surface and Coatings Technology*, vol. 140, 175–181, 2001

[SPA 81] T. Spalvins, B. Buzek, "Frictional and morphological characteristics of ion-plated soft metallic films", *Thin Solid Films*, vol. 84, 267–272, 1981

[SPR 80] J.M. Sprauel, Détermination des constantes élastiques radiocristallographiques d'un acier inoxydable austénitique, PhD Thesis, Université de Paris Sud, centre d'Orsay, 1980

[STA 04] M.M. Stack, B.D. Jana, "Modelling particulate erosion-corrosion in aqueous slurries: some views on the construction of erosion-corrosion maps for a range of pure metals", *Wear*, vol. 256(9–10), 986–1004, 2004

[STA 06] M.M. Stack, M.M. Antonov, I. Hussainovo, "Some views on the erosioncorrosion response of bulk chromium carbide based cermets", *Journal of Physics D: Applied Physics*, vol. 39, 3165–3174, 2006

[STE 85] P.A. Steinmann, H.E. Hinterman, "Adhesion of TiC ans TiCN coatings on steel", *Journal of Vacuum Science and Technoloy*, NO.A3, 2394–4000, 1985

[STE 87] P.A. Steinmann, Y. Tardy, H.E. Hinterman, "Adhesion testing by the scratch test method: the influence of intrinsic and extrinsic parameters on the critical load", *Thin Solid Films*, vol. 154, 333–349, 1987

[STI 04] D. Stievenard, "Microscope à effet tunnel", in M. Lehman, C. Dupras and P. Houdy (ed.), *Les nanosciences*, Belin, Paris, 72–93, 2004

[STO 09] G.G. Stoney, "The tension of metallic films deposited by electrolysis", *Proceedings of the Royal Society of London*, Series A, vol 82, 172–175, 1909.

[TAB 51] D. Tabor, *Hardness of Metals*, Oxford University Press, London, 1951

[TAK 85] J. Takadoum, J.C. Pivin, J. Chaumont, C. Roques-Carmes, "Friction and wear of amorphous Ni-B, Ni-P films obtained by ion implantation into nickel", *Journal of Materials Science*, vol. 20, 1480–1493, 1985

[TAK 86] J. Takadoum, Corrélation des propriétés rhéologiques et de réactivité au comportement tribologique des alliages Ni1-xBx et Ni1-xPx cristallins et amorphes, élaborés par implantation ionique, PhD Thesis, University of Paris VI, Orsay, 1986

[TAK 87] J. Takadoum, J.C. Pivin, H.M. Pollock, J.D.C. Ross, H. Bernas, "The mechanical properties of boron- and phosphorus-implanted nickel, discussed in terms of increasing disorder and amorphicity", *Nuclear Instruments and Methods in Physics Research B*, vol. 18, 153–160, 1987

[TAK 92] J. Takadoum, C. Roques-Carmes, "Influence of the oxidation activity of metals on friction and wear of ceramic–metal systems", *Surface and Coatings Technology*, vol. 52(2), 153–158, 1992

[TAK 93a] J. Takadoum, "Tribological behaviour of alumina sliding on several kinds of materials", *Wear*, vol. 170, 285–290, 1993

[TAK 93b] J. Takadoum, Z. Zsiga, "Effect of water vapour in air on friction and wear of Al2O3, Si3N4 and partially stabilized zirconia sliding on various metals", *Journal of Materials Science Letters*, vol. 12, 791–793, 1993

[TAK 93c] J. Takadoum, "Tribological behaviour of alumina sliding on several kinds of materials", *Wear*, vol. 170, 285–290, 1993

[TAK 94a] J. Takadoum, Z. Zsiga, C. Roques-Carmes, "Wear mechanism of silicon carbide: new observations", *Wear*, vol. 174, 239–242, 1994

[TAK 94b] J. Takadoum, Z. Zsiga, M. Ben Rhouma, C. Roques-Carmes, "Correlation between friction coefficient and wear mechanism of SiC/SiC system", *Journal of Materials Science Letters*, vol. 13(7), 474–476, 1994

[TAK 96] J. Takadoum, "The influence of potential on the tribocorrosion of nickel and iron in sulfuric acid solution", *Corrosion Science*, vol. 38(4), 643–654, 1996

[TAK 97a] J. Takadoum, H. Houmid Bennani, "Influence of substrate roughness and coating thickness on adhesion, friction and wear of TiN films", *Surface and Coatings Technology*, vol. 96(2–3), 272–282, 1997

[TAK 97b] J. Takadoum, H. Houmid Bennani, M. Allouard, "Friction and wear characteristics of TiN, TiCN and diamond-like carbon films", *Surface and Coatings Technology*, vol. 88(1–3), 232–238, 1997

[TAK 97c] J. Takadoum, "Etude de la tribocorrosion du nickel et de l'acier inoxydable 316 en milieu H_2SO_4", *Matériaux et Techniques*, numéro spécial Corrosion Tribocorrosion, p. 29–31, July 1997

[TAK 01] J. Takadoum, B. Cretin, "Study of adhesion and tribological properties of some ceramic films" in Mittal K. L. (ed.), *Adhesion Aspects of Thin Films*, VSP editions, Netherlands, p. 195–205, 2001

[TAK 03] J. Takadoum, J.Y. Rauch, J.M. Cattenot, N. Martin, "Comparative study of mechanical and tribological properties of CNx and DLC films deposited by PECVD technique", *Surface and Coatings Technology*, vol. 174–175, 427–433, 2003

[TAZ 78] M. Tazaki, M. Nishibori, K. Kinosita, "Ultra-microhardness of vacuum-deposited films. II: results for silver, gold, copper, MgF2, LiF and ZnS", *Thin Solid Films*, vol. 51, p.13-21, 1978

[TER 96] P. Terrat, J.J. Reymond, *Le Vide*, no. 280, p. 219–224, April-May-June 1996

[THO 98] M.D. Thouless, "An analysis of spalling in the microscratch test", *Engineering Fracture Mechanics*, vol. 61, 75–81, 1998

[THOB 99] A. Thobor, C. Rousselot, J. Takadoum, N. Martin, *Le Vide, Science, Technique et Applications*, vol. 291, 138, 1999

[THOB 00] A. Thobor, C. Rousselot, C. Clément, J. Takadoum, N. Martin, R. Sanjines, F. Lévy, "Enhancement of mechanical properties of TiN/AlN multilayers by modifying the number and the quality of interfaces", *Surface and Coatings Technology*, vol. 124, 210–221, 2000

[THOMAST 05] T. Thomas, B.-G. Rosen, H. Zahouani (ed.), *Proceedings of the 10th International Conference on Metrology and Properties of Engineering Surfaces*, Saint-Etienne University, 4–7 July 2005

[THOMASTR 99] T.R. Thomas, *Rough Surfaces*, Imperial College Press, London, 2nd edition, 1999.

[THOR 74] J.A. Thoronton, *J. Vac. Sci. Technol.*, vol 11(4), 666–670, 1974

[TRO 82] J.P. Trotignon, J. Verdu, M. Piperaud, A. Dobraczynski, *Précis de matières plastiques*, Nathan, Paris, 5th edition, 1982

[VAL 86] J.A. Valli, "A review of adhesion test methods for thin hard coatings", *Journal of Vacuum Science and Technology*, vol. A4, 3001–3014, 1986

[VANO 88a] C.J. Van Oss, R.J. Good, M.K. Chaudhury, "Additive and non additive surface tension components and the interpretation of contact angles", *Langmuir*, vol. 4, 884–891, 1988

[VANO 88b] C.J. Van Oss, M.K. Chaudhury, R.J. Good, "Interfacial Lifshitz-van der Walls and polar interactions in macroscopic systems", *Chemical Reviews*, vol. 88, 927– 941, 1988

[VANP 04] A.C. Van Popta, J.C. Sita, J.M. Brett, "Optical properties of porous helical thin films and the effects of postdeposition annealing", *SPIE*, vol. 5464, 198–208, 2004

[VANP 05] A.C. Van Popta, M.M. Hawkeye, J.C. Sit, M.J. Brett, *Nanotechnology*, vol. 4, 269, 2005

[VAZ 04] F. Vaz, P. Cerqueira, L. Rebouta, S.M.C. Nascimento, E. Alves, P. Goudeau, J.P. Rivière, K. Pischow, L. De Rijk, "Structural, optical and mechanical properties of coloured TiNxOy thin films", *Thin Solid Films*, vol. 447–448, 449–454, 2004

[VEP 95] S. Veprek, S. Reiprich, S. Li, "Superhard nanocrystalline composite materials: the TiN/Si_3N_4 system", *Applied Physics Letters*, vol. 66(20), 2640–2642, 1995

[VEP 00] S. Veprek, A. Niederhofer, K. Moto, T. Bolom, H.D. Mannling, P. Nesladek, G. Dollinger, A. Bergmaier, "Composition, nanostructure and origin of the ultrahardness in nc–TiN/a–Si_3N_4/a– and nc–$TiSi_2$ nanocomposites with HV=80 to ≥105 GPa", *Surface and Coatings Technology*, vol. 133–134, 152, 2000

[VOE 96] A.A. Voevodin, J.M. Schneider, C. Rebholz, A. Matthews, "Multilayer composite ceramicmetal–DLC coatings for sliding wear applications", *Tribology International*, vol. 29(7), 559–570, 1996

[WAL 06] J. Wall, H. Choo, T.N. Tiegs, P.K. Liaw, "Thermal residual stress evolution in a TiC–50 vol.% Ni_3Al cermet", *Materials Science and Engineering: A*, vol. 421(1–2), 40–45, 2006

[WANG 06] Y. Wang, S. Lim, J.L. Luo, Z.H. Xu, "Tribological and corrosion behaviors of Al_2O_3/polymer nanocomposite coatings", *Wear*, vol. 260(9–10), 976–983, 2006

[WANS 99] O. Wänstrand, M. Larsson, P. Hedenqvist, "Mechanical and tribological evaluation of PVD WC/C coatings", *Surface and Coatings Technology*, vol. 111(2–3), 247–254,1999

[WEI 82] C. Weissmantel, K. Bewilogua, K. Breuer, D. Dietrich, U. Ebersbach, H. –J. Erler, B. Rau and G. Reisse, "Preparation and properties of hard i–C and i–BN coatings", *Thin Solid Films*, vol. 96, 31–44, 1982.

[WEI 97] Z.Q.X. Wei, Z. Xushou, "Micro structure and tribological characteristic of Ni-Mo multilayer film deposited by ion beam mixing", *Surface and Coatings Technology*, vol. 91(1–2), 50–56, 1997

[WEN 95] E.J. Wentzel, C. Allen, "Erosion–corrosion resistance of tungsten carbide hard metals with different binder compositions", *Wear*, vol. 181–183, 63–69, 1995

[WIK 01] U. Wiklund, S. Hogmark, J. Gunnars, "Delamination of thin hard coatings induced by combined residual stress and topography" in Mittal K. L. (ed.), *Adhesion Aspects of Thin Films*, VSP editions, The Netherlands, p. 51–65, 2001

[WIL 94] J.A. Williams, *Engineering Tribology*, Oxford University Press, Oxford, 1994

[YAM 04] Y. Yamamoto, M. Hashimoto, "Friction and wear of water lubricated PEEK and PPS sliding contacts: Part 2. Composites with carbon or glass fibre", *Wear*, vol. 257, 181–189, 2004

[YAO 97] H. Yao, Y.L. Su, "The tribological potential of CrN and Cr(C,N) deposited by multi-arc PVD process", *Wear*, vol. 212(1), 85–94, 1997

[YON 06] M. Yonggang, H. Bo, C. Qiuying, "Control of local friction of metal/ceramic contacts in aqueous solutions with an electrochemical method", *Wear*, vol. 260(3), 305–309, 2006

[YUG 95] Z. Yugui, Y. Zhiming, Y.W. Xiang, K. Wei, "The synergistic effect between erosion and corrosion in acidic slurry medium", *Wear* (186–187), 555–561, 1995

[ZHANGG 06] G. Zhang, H. Liao, H. Lia, C. Mateus, J.M. Bordes, C. Coddet, "On dry sliding friction and wear behaviour of PEEK and PEEK/SiC-composite coatings", *Wear*, vol. 260, 594–600, 2006

[ZHANGS 97] S.W. Zhang, "State of the art of polymer tribology", in I. M. Hutchings (ed.), *New Directions in Tribology*, Mechanical Engineering Publications, MEP, London, p. 467–481, 1997

[ZHANGT 94] T.C. Zhang, X.X. Jiang, S.Z. Li, X.C. Lu, "A quantitative estimation of the synergy between corrosion and abrasion", *Corrosion Science*, vol. 36(12), 1953–1962, 1994

[ZUM 96] K.H. Zum Gahr, "Modeling and microstructural modification of alumina ceramic for improved tribological properties", *Wear*, vol. 200(1–2), 215–224, 1996

[ZUM 00] K.H. Zum Gahr, J. Schneider, "Surface modification of ceramics for improved tribological properties", *Ceramic International*, vol. 26(4), 363–370, 2000

[ZYG 89] R. Zugmunt, *Tribology of Miniature Systems*, Tribology series, Elsevier, no. 13, p. 20, 1989

Index

A
Abbott curve, 7-8
abrasive wear, 83-84, 111-112, 116-117, 134, 144, 166
accommodation, 87-88
additives, 88-89, 93-95, 116
adherence, 178, 191-194, 197-199
adhesion, 17-21, 63, 65, 70, 75, 77-81, 88, 110, 117, 131, 134, 146, 157-159, 172-173, 188-191
adhesion energy, 19, 21, 191
adhesive contact, 59, 62, 66, 130
adhesive force, 60, 62, 71, 75-77, 80-81
adhesive wear, 83-84, 111, 117
alloy, 30, 98, 108-113, 124, 127, 134, 136, 140, 142, 147, 154, 159, 171
alumina (aluminum oxide), 12, 20, 25, 30-31, 36, 123-127, 130-131, 135-137, 168
aluminum, 10, 25, 36, 101, 104-105, 112-113, 123, 134-137, 154, 167, 171-172, 177, 185
amorphization, 64, 168
amorphous, 81-82, 154, 167
anodic oxidation, 134-136
atomic force microscope (AFM), 14-17, 75-79

B
beam, 10-16, 38, 43-45, 127, 137-138, 142, 144, 157, 159
bearing, 7-8, 87, 91-92, 113, 123
Berkovich, 31, 34
blister test, 192-193
boronizing, 143
Brinell, 24, 26
brittleness, 24, 34, 108, 137, 149

C
capillary force, 17, 71-71, 80
carbon, 2, 30, 47, 75-76, 81-82, 95, 112, 120-121, 143, 171-172, 197-198, 203
carbon nitride, 143, 172, 174, 177-178
carburizing, 143, 178
ceramic, 25, 34-35, 66, 99, 100, 110, 112-113, 120-133, 142, 144, 147, 152-155, 170, 173, 192, 198, 204
cermet, 131-133
coating, 15, 44-47, 97-98, 104, 110, 112, 116-117, 133, 146-155, 158-159, 165-204
coefficient of friction, 57, 62-63, 66, 100
cohesion, 18, 20, 35

composite, 95-96, 110, 120-121, 124, 131, 133, 152-154, 165-168, 173, 182-189, 201
copper, 12, 15, 20, 25, 30, 36, 68, 103, 113, 121, 154, 169-171, 185, 191
corrosion, 2, 44-45, 82, 94, 98-108, 111-113, 123, 134, 142, 149, 169
crack, 24, 35-37, 57, 83, 117, 123-127, 130-132, 146, 173, 186, 191-200
cubic boron, 122, 124

D

dangling bonds, 2, 18
deposit, 15, 45, 95-98, 127, 142-143, 147-167, 170-173, 177, 183-184, 190-192, 197-199, 201-204
deposition, 97-98, 112, 133, 142, 146, 148-167, 170, 175, 190-191, 201
detergent, 89, 94
diffusion, 46, 64, 85, 142-143, 146, 172-173, 191
Dupré, 191

E

elastomer, 27
electrochemical, 97, 100-102, 105-106, 148-149, 152, 154, 201
electrodeposition, 47, 148
electrostatic force, 71, 76, 191
erosion-corrosion, 98-99, 106

F

fatigue wear, 83-84, 112, 117
fracture propagation, 34-37
fracture toughness, 34-36, 122-126, 132
fragility, 121, 124
friction coefficient, 57, 64-69, 91, 95-98, 103-104, 113-121, 126-133, 140, 152, 168-174, 195
friction materials, 89, 94, 121

G

glancing angle deposition (GLAD), 161, 164-166
graphite, 66, 94-98, 112-113, 120-121, 171, 174
greases, 47, 89, 94, 116

H

hardness, 2, 24-27, 30, 36-37, 59, 82, 86, 110-113, 116-118, 121-126, 131-133, 136, 140, 143, 145, 150-154, 166-167, 170-173, 176-189, 195, 199-200
Hertz, 53, 55, 59-61, 194
humidity, 66, 71, 76, 95, 109, 127-130, 165, 171
hydration, 76
hydrocarbon, 2, 92
hydrodynamic, 66, 91-92
hydroxide, 101, 127-130, 134, 168

I

indentation, 25-28, 31-33, 36-37, 118-119, 137, 178-192, 199-201
  indentation size effect (ISE), 179
indenter, 12, 24-27, 30-34, 141, 179-183, 187, 194-195, 199
inhibitor, 94
ion implantation, 118, 127, 137, 140
iron, 25, 30, 36, 46, 101, 103, 107-108-113, 121, 124, 143, 153, 171, 177, 185
isoelectric point (IEP), 77

K, L

kurtosis, 6
lamellar solids, 94
lubricant, 47, 66, 83, 86-99, 103, 109-113, 116, 128, 137, 152, 168, 171
lubricating, 91-92, 95, 98, 120, 130, 172
lubrication, 49, 56, 89-94, 112-113, 146, 167

M

micromanipulation, 79-80
mineral oil, 92-93
molybdenum disulfide, 94-98

N

nanofriction, 81
nano-indentation, 30-33, 118-119, 165-166, 179
nanoparticles, 167
nanotribology, 71, 77, 81

O

oxidation, 20, 64, 83, 85, 93, 99, 104-105, 110, 112, 124, 134-136, 148, 154, 167-168, 201
oxide, 3, 12, 20, 25, 30-31, 36, 44-45, 77-79, 83, 94, 98, 101-105, 112, 122-123, 127-131, 134, 137, 143, 155-159, 167-168, 174-179

P

passivation, 2, 45, 101-106, 134
peeling, 192-193
physical vapor deposition (PVD), 97, 154, 156, 164-168, 171-172, 175, 180, 198, 201, 203
plasma, 98, 147, 155-156
    plasma assisted chemical vapor deposition (PACVD), 155-156, 166
    plasma enhanced chemical vapor deposition (PECVD), 155, 167, 171
    plasma torch, 147
plasticity, 49, 59
ploughing, 70
polyacetal, 116, 121
polyamide, 20, 114-120
polyetheretherketone (PEEK), 114-117, 120-121
polyimide, 88, 114-116
polymer, 25, 27, 45, 47, 66, 70, 82, 88-89, 110, 113-120, 140, 191-192
polymethacrylates, 89

polyoxymethylene (POM), 114-116
polytetrafluoroethylene (PTFE), 20, 66, 88, 95-96, 100, 114-116, 120-121, 152
porosity, 130, 161, 165-166, 171, 192
pulsed signals, 137
PV product, 88

R

residual stresses, 3, 24, 37-38, 133, 144-146, 195, 201-202
Rockwell, 24-27, 194
roughness, 4-10, 69-70, 91, 117-118, 146, 161, 166, 171, 195-198

S

Society of Automotive Engineers (SAE), 90-91, 185
scratch test, 194-199
shear, 50-52, 56, 65, 68, 70, 83, 87-88, 94-95, 145
Shore, 24-27
shot peening, 37, 144-146, 178
silicon carbide, 25, 30, 36, 86, 121-124, 128, 153
silicon nitride, 14, 30, 77-79, 103, 123, 128
skewness (sk), 6, 7, 10
sliding, 57, 62-68, 81-82, 85-86, 93-98, 103-106, 109, 126, 129, 166, 168, 194-198
sputtering, 15, 46, 95, 157-161, 170-172, 175-177
steel, 27, 46, 68-69, 84-85, 88, 97-98, 104-106, 110-112, 117-118, 121-124, 140, 143-145, 151-155, 159, 169-174, 178, 184-185, 191, 197-198, 203
scanning tunneling microscope (STM), 13-14
Stoney, 201-203
Stribeck curve, 91
superalloy, 112
surface energy, 2, 18-24, 35, 61-62, 74-75, 110, 113, 118, 133

surface treatment, 37, 111, 118, 127, 133-134, 138, 143-146, 154, 178
synthetic oil, 92-94

T
thermal projection, 147
thermo-mechanical treatment, 111
thermoplastic, 27
third body, 86-88, 109
titanium dioxide, 176-177
titanium nitride, 112, 167, 169, 176-177
toughness, 24, 34-36, 111, 116, 122-126, 132, 166, 200
tribochemical wear, 83-84
tribocorrosion, 88, 98-103
tribometer, 66-67, 100, 129, 132, 168
tribo-oxidation, 83

U, V
Van der Waals force, 17, 19, 23, 59, 71-76, 80
varnishes, 47, 94-98, 192
vegetable oils, 93
Vickers, 24-26, 31, 36-37, 122, 137, 143, 183-187, 199, 201
viscosity index, 90, 94

W, X, Y, Z
wear-corrosion, 82, 98
x-ray, 38-45, 202
yield criteria, 52
zirconia, 123-127, 152, 178